Insect-Plant Interactions

Volume I

Editor

Elizabeth A. Bernays, Ph.D.
Professor of Entomology and
Professor of Zoology
University of California, Berkeley
Berkeley, California

CRC Press, Inc.
Boca Raton, Florida

Library of Congress Cataloging-in-Publication Data

Insect-plant interactions/editor, Elizabeth A. Bernays.
p. cm.
Bibliography: p.
Includes index.
ISBN 0-8493-4121-3 (v. I)
1. Insect-plant relationships. 2. Insect pests. I. Bernays, E.
A. (Elizabeth A.)
SB931.I54 1989
632'.7—dc19 88-35918
CIP

Direct all inquiries to CRC Press, Inc., 2000 Corporate Blvd., N.W., Boca Raton, Florida, 33431.

International Standard Book Number 0-8493-4121-3 (Volume I)
Library of Congress Number 88-35918

Printed in the United States

PREFACE

Insect-Plant Interactions is a selection of reviews with interest to researchers in a variety of disciplines. The articles are on topics with considerable current interest and include recent work with potential for altering research trends.

Campbell reviews the role of endosymbionts in insect herbivores, and in particular the possibilities for future development of the field and the potential importance of the microorganisms involved, in the adaptation of insects to plants.

Crawley discusses the relative importance of vertebrates and insects in herbivory—a topic of considerable interest to those concerned with selection pressure by herbivores on plants and the evolution of plant defenses.

Riemer and Whittaker have brought together a very scattered literature on the significance of aerial pollution for insect herbivory. This field is of rapidly increasing importance in today's world and the possible routes for eventual effects on insect herbivores are many. Although conclusions can be only tentative, this article presents the way to more useful research.

Waterman and Mole review production of plant secondary compounds from a plant physiological standpoint. This is of particular importance today because, after decades of having mainly the zoologists' theories of plant chemical defense, a balanced perspective is being realized.

Welter shows that the different approaches to the topic of insect damage and its effects on plant photosynthesis lead to different conclusions. This review puts together the different approaches in a comprehensive manner allowing a more general conclusion to be reached.

THE EDITOR

Dr. Elizabeth Bernays is Professor of Entomology and Professor of Zoology at the University of California, Berkeley. After receiving a B.S. with honors in 1962 at the University of Queensland, Brisbane, Australia, she travelled in Europe and taught biology, before studying for the M.Sc. and then Ph.D. at the University of London, England. Her Ph.D. was a study of the hatching mechanism in the desert locust involving morphology, physiology, and behavior.

From 1970 to 1983, Dr. Bernays was scientist, senior scientist, and then principal scientist in a British Government Research Institute involved with research on insect pests in developing countries. This institute is now called the Overseas Development Natural Resources Institute. During this period, she worked in the laboratory on the physiological regulation of feeding behavior in grasshoppers and locusts, and on the effects of various plant compounds on behavior and physiology of plant-feeding insects. In the field, she worked on cassava pest biology in Nigeria and cereal resistance to insects in India. With Dr. R. F. Chapman she organized, in 1978, the Fourth International Symposium on Plant-Insect Interactions and edited the resultant book on that topic.

In 1983, Dr. Bernays moved to Berkeley, where she works on a variety of topics in plant-insect interactions. She is in the Department of Entomology, Division of Biological Control and the Department of Zoology.

In the field of plant-insect interactions, Dr. Bernays combines her physiological and behavioral experience with functional morphology and ecology, to provide a multidisciplinary approach. Her major interests concern host range in phytophagous insects: functional and causal mechanisms of host choice; variation and the role of learning; chemical ecology of plants and insects; biological control of weeds.

Dr. Bernays has published widely in international journals of ecology, morphology, behavior, and physiology and is an assistant editor for the *Journal of Chemical Ecology, Journal of Insect Behavior,* and *Entomologia Experimentalis et Applicata.* She has presented numerous papers in universities and institutes in the U.S. and Europe.

TABLE OF CONTENTS

1 On the Role of Microbial Symbiotes in Herbivorous Insects

Bruce C. Campbell
U.S. Department of Agriculture-Agricultural Research Service
Western Regional Research Center
Albany, California

TABLE OF CONTENTS

I. INTRODUCTION

For almost a century it has been known that many insects, across a broad taxonomic range, have an obligate association with prokaryotic and/or eukaryotic microorganisms which are harbored extra- or intracellularly. The most comprehensive review on the subject involving these symbiotes in insects was written by Buchner more than 30 years ago.[79] Whereas much is known about the taxonomy and biochemistry of the extracellular symbiotes, a corresponding understanding of the biology of intracellular symbiotes (i.e., endosymbiotes) has yet to be achieved. This is primarily because insect endosymbiotes have never been successfully cultured outside their host. However, recent efforts in the study of insect endosymbiotes have employed genetic engineering techniques that permit the study of genes and/or gene products which designate the taxonomy and biochemistry of the endosymbiotes.

This chapter presents information on the role of symbiotes in insect-plant interactions, from a biochemical perspective, with emphasis on the endosymbiotes. A consideration of the role of insect symbiotes in the ecology of insect-plant interactions has already been evaluated by Jones.[293] Unlike that of extracellular symbiotes, it is difficult to come to definitive conclusions as to the role of endosymbiotes in the interaction of insects with their host-plants without a thorough understanding of their biochemistry. There is a large body of information that considers the biochemistry of insect endosymbiotes indirectly. Studies have compared the biosynthetic products and/or physiological performances between insects that possess endosymbiotes with those whose symbiotes have been eliminated by antibiotics or some other treatment (e.g., heat). However, because of the potential of these treatments to affect host tissue, one should be cautious with the conclusions drawn from such studies. The available knowledge on extracellular symbiotes may be relevant to our understanding of the biological role of insect endosymbiotes. Hence, some of the pertinent literature on these extracellular symbiotes is reviewed prior to that of insect endosymbiotes.

II. EXTRACELLULAR SYMBIOTES

Extracellular symbiotes between microorganisms and insects has developed both internally and externally (viz., extracuticularly) in insects. By far, the most widely studied extracellular symbiotes are the protozoa that inhabit the hindgut of termites.[120,121,262,263,568] Additionally, there are numerous reviews which emphasize the literature on extracellular microorganisms that inhabit insects in general.[47,162,238,521-524,536] There are other less well-defined examples of extracellular microorganisms inhabiting the alimentary tract of insects (e.g., bacteria of the esophageal bulb in tephritid larvae) or residing on the cuticle (e.g., saprophytic or mycangial fungi of wasps and/or xylophagous beetles, respectively).

A. INTERNAL EXTRACELLULAR SYMBIOTES

1. Termites

Reviews are available on the biochemical,[60,410,442] taxonomic,[43,59,228] and overall biological roles of termite symbiotes.[325,326] The most widely studied and best-defined role of termite symbiotic microbiota is by far their production of cellulolytic enzymes for the digestion of wood. Almost 50 years ago, it was discovered that certain protozoa in the hindgut of *Zootermopsis* produced enzymes capable of fermenting cellulose.[262,263,265] More-recent studies have greatly expanded the known number of taxa where cellulase, carboxymethylcellulase, or cellobiase activities in termites are produced either by protozoan flagellates or bacteria inhabiting the hindgut.[248,387,451,532,533,569,571] Some termites (e.g., *Nastutitermes* and *Coptotermes*) are capable of producing their own cellulolytic enzymes in addition to those supplemented by bacterial symbiotes.[248] In many cases, the kinetic and physical characteristics of termite cellulolytic enzymes have been defined.[465,466] It has also been suggested that the symbiotic microflora of termites produce ligninases[83] and chitinases.[548]

The above-mentioned studies establish that one of the biological roles of termite symbiotes is to produce enzymes which digest wood and/or cultivated fungi. Such enzymes enable subterranean termites to meet carbon-source requirements. However, numerous studies also show that these symbiotes play a second, vital role in the synthesis of termite nutritional requisites. Hungate suggested that termite-gut symbiotes were involved in fixation of atmospheric nitrogen.[264] Recent studies confirm that nitrogen is fixed by bacteria in the hindgut of many species of termites.[62,196,430,460,469,470] Moreover, the nutritional status of termite colonies is dependent upon the bioavailability of a nitrogen source from either fixed nitrogen or from the metabolism of uric acid by gut symbiotes.[32,352,461-464] In addition to nitrogen metabolism, termite-gut symbiotes play a role in the synthesis and metabolism of fatty acids.[247,384,443] Lipid metabolism by symbiotic bacteria in termites is also involved in the production of precursors for synthesis of amino acids.[385] The symbiotic relationship in termites is not limited to termite-gut symbiote interactions. Certain hindgut symbiotes in termites interact interdependently in the synthesis and metabolism of lactates.[497] Bacterial symbiotes in termites maintain a specific redox potential in the hindgut that inhibits the growth of nonsymbiotic or potentially pathogenic bacteria.[546] The symbiotes may also be involved in the synthesis of noxious chemicals used in defense of the colony against ants or termite predators[468] and contribute to atmospheric methane.[477]

The ability to culture and describe physical characteristics of termite symbiotes outside of their host has greatly facilitated their definitive identification and classification. Many of the protozoan symbiotes[61,249,343,570] and bacterial symbiotes have been classified.[37,39,137,183,184,343,496,534] The specialized morphology and epithelial tissue of the alimentary tract of termites which enables the attachment or establishment of symbiotic microflora has also been noted.[38,40,46]

2. Other Insects

The symbiotic microflora of the alimentary tract of a variety of other xylophagous insects (e.g., wood-inhabiting cockroaches and beetles) has been studied almost to the same extent as that of termites. To a lesser degree, the biological role of the extracellular gut symbiotes of certain Diptera (e.g., larval symbiotes of various tephritids and the screwworm fly), Lepidoptera, and chrysopids have been studied as well. However, in many cases with these latter insects, the contribution that the symbiote provides towards host survival is not clear.

a. Cockroaches

As with termites, one of the well-established roles of the microbial community in cockroach intestines is the biosynthesis of enzymes for the decomposition of cellulose.[43,123,132,216,570] Furthermore, intestinal symbiotes also play a significant role in nitrogen recycling in cockroaches.[124] They are involved in the metabolism of uric acid and the overall excretion of nitrogen.[155,412,413] By synthesizing chitinases, symbiotes assist in the retrieval of nitrogen from chitin.[548] Symbiotes may also play a role in the biosynthesis of nutritionally important polyunsaturated fatty acids in cockroaches.[247] The significance of this biosynthesis to cockroach nutrition is equivocal, however, since certain roaches can synthesize some of these fatty acids de novo.[42] The ability to culture and/or study cockroach symbiotes outside the host has also facilitated their systematic classification. A number of studies have characterized the bacterial symbiotes of cockroaches[35,52,133-135,189,190] as well as the protozoan symbiotes.[43,52,64,570] The extracellular nature of these symbiotes has also enabled examination of specializations in their morphology which secures them in the gut of cockroaches.[36,51,136] Cockroach symbiotes are generally transmitted transovarially during oogenesis.[489] In addition to the extracellular symbiotes, cockroaches possess endosymbiotes (*vide infra*).

b. Coleoptera

Symbiotic microorganisms of beetles appear to be the chief producers of an array of digestive enzymes which are capable of the hydrolytic breakdown of plant matrix polysaccharides. In many cases these enzymes are ingested from fungal symbiotes growing outside the host (*vide infra*).[374] However, extracellular gut symbiotes provide various polysaccharases which enable xylophagous beetles to digest wood, leaf litter, or detritus. Moreover, the provision of these enzymes by symbiotes plays an overall significant role in the ecophysiology of a number of beetles which are forest pests.[106,149] Chararas and colleagues at l'Institut National Agronomique in Paris have published numerous studies on the wood-digesting enzymes of various xylophagous beetles.[106-114,116-118,127-129,340] They found that various species of xylophagous beetles (e.g., scolytids, cerambycids, pyrochroidids, and buprestids) possess a wide range of enzymes capable of hydrolyzing plant polysaccharides that are refractory to digestive enzymes normally produced by epithelial cells of invertebrates. These polysaccharide substrates include cellulose, xylan, arabinoxylan and other hemicelluloses, lignin, lichenan, laminarin, carboxymethylcellulose, pectin, and a variety of other β-linked phenyl and methyl glycosides. Their work also indicates that the presence of these various polysaccharases in the alimentary tract of xylophagous beetles is either through acquisition from saprophytic fungi,[117,128] biosynthesis from gut symbiotes.[111,112,116-118,128,129,340] or synthesis by gut epithelium.[107,110,113,127-129] The breakdown of these polysaccharides releases free monosaccharides from plant constituents which are ordinarily nutritionally inert. These monosaccharides are eventually assimilated and metabolized, thus providing the host insect with a source of energy. The hindgut of some scarabeids has been modified into a fermentation chamber which contains bacteria and/or protozoa capable of digesting cellulose.[28,30,524,551] However, the major contribution of the studies by Chararas and colleagues is the idea that variation in the hydrolytic kinetics of different polysaccharases can be a factor which dictates compatibility between xylophagous beetles and their host trees. Furthermore, some of their studies are the only reports wherein endosymbiotes are indirectly shown to synthesize enzymes able to hydrolyze plant matrix heteropolysaccharides. This topic is covered in further detail in the discussion of the role of endosymbiotes in insects, (*vide infra*).

In addition to the production of wood-digesting enzymes, extracellular gut symbiotes of beetles are involved in various other metabolic, physiological, or behavioral functions of the beetle. Chararas et al. have shown that bacteria in one scolytid species, *Phoracantha semipunctata,* can synthesize B-group vitamins.[114] Bacteria in the alimentary tract of some other species of scolytids are able to decompose terpenoids, thereby alleviating toxicity.[115] Phenol in the colleterial gland of the New Zealand grass grub is believed to be produced by bacteria.[261] Brand et al. found that bacteria present in the gut of *Ips paraconfusus* could convert α-pinene into the pheromones *cis-* and *trans-*verbenol.[120] Other studies of the bacterial symbiotes of bark beetles have successfully characterized the microbial flora at various life stages and noted their ability to fix atomospheric nitrogen.[63,81] Gut symbiotes of scarab beetles play a role in the biosynthesis of cellulases,[27,28,30] production of methane,[28,29] and the biosynthesis of fatty acids.[28] However, cellulase production by the phytophagous coccinellid, *Epilachna varivestis*, is thought not to be derived from symbiotic microflora.[531] Yeast-like symbiotes in the midgut of various cerambycids may facilitate nitrogen utilization and excretion.[524] Mycangial and saprophytic fungi also play an important role in the ecophysiology of xylophagous beetles. Since these fungi are harbored on the surface of the exoskeleton, their roles are discussed in Section II.B.

Lastly, a unique role of prokaryotic symbiotes is to induce nongenetic reproductive isolation between different strains of flour beetles.[547] This discovery has major implications in our understanding of speciation and biotype formation in any insect that harbors bacterial symbiotes. The role of symbiotes in the evolution of host-race formation in insects is discussed further in Sections II.A.c and III.

c. Diptera

The role of extracellular symbiotes associated with the alimentary tract of dipteran larvae and/or adults is only partially understood at present. Most of the research on the functional role of dipteran symbiotes has involved the frugivorous tephritids. Petri was the first to discover that a tephritid, the olive fruit fly (*Davus oleae*), always possesses certain microorganisms believed to be symbiotes.[452,453] Later, it was learned that another tephritid, the apple maggot (*Rhagoletis pomonella*), also harbors bacteria believed to cause the rotting of fruit associated with infestations by the larvae.[4] It is now well established that trypetines and dacines all possess a specialized esophageal bulb, or cephalic vesicle, for the containment of one or more species of bacteria.[21,49,210,290,456,478,479,538,572] Elimination of these symbiotic bacteria by treatment with antibiotics generally results in the disruption of larval development and/or adult fecundity,[4-6,198,199,229,231,232,539] but this functional role in fruit fly survival is still unclear. In some studies, it has been shown that symbiotic bacteria or other bacteria associated in one way or another with fruit flies are capable of providing the host insect with certain nutritional requisites or amino acids.[229,232,407] However, a role for these bacterial symbiotes to provide essential nutrients absent from fruit is probably more questionable in view of the fact that symbiotes also occur in tephritines whose diet is inherently nutritionally complete (e.g., tephritinae feeding on composites).[209] It has been shown that a bacterial symbiote, *Pseudomonas melophthora*, of the apple maggot is able to metabolically degrade several pesticides *in vitro*.[48] Symbiotic bacteria may play a role in synthesis of oviposition stimulants[251] or in the production of various nutrients in aphid honeydew[232] for adult tephritids. It has been shown that leaf-surface bacteria are a direct source of food for *D. tryons,* the Queensland fruit fly.[126] Whether the bacteria isolated from the fruit flies are true symbiotes remains a question. They may be merely facultatively or opportunistically associated with either the larvae or the adult flies.[147,159] Some of the bacteria isolated from leaf surfaces and fruits infested by fruit flies can also be isolated from the alimentary tract and eggs of fruit flies.[159,187]

However, two well-devised studies on the biology of tephritid symbiotes recently identified some of the bacterial symbiotes of *R. pomonella* and provided some ideas of their role.[259,488] Rossiter et al. found that the *Klebsiella oxytoca* and, tentatively, *Enterobacter cloacae* were the only two species of bacteria that persisted through all lifestages of *R. pomonella*.[488] They suggested that, in view of the fact that *K. oxytoca* was known to produce pectinolytic enzymes, the symbiotic bacteria played a role in the preliminary degradation of fruit tissue, thus eventually facilitating digestion of fruit by larvae. The question of whether different strains of symbiotic bacteria play a role in the establishment of fruit fly host-races (viz., biotypes) was addressed by Howard et al.[259] They found no basis for a conclusion that different species of bacteria or strains of *K.*

oxytoca are associated either with different species or biotypes of *Rhagoletis*. They also concluded that there are no obligate relationships between the *Rhagoletis* species in their study and any particular species or strain of bacteria. Moreover, none of the bacteria isolated from the flies could fix atmospheric nitrogen at a detectable level. Similarly, Fitt and O'Brien found a diversity of bacteria in four species of *Dacus*.[187] However, they could not establish the existence of any strict bacteria-fly associations. Some of the bacteria could potentially serve a nutritional role or provide proteolytic enzymes, but they concluded that the bacteria-fly symbiosis was simply fortuitous.

Although most of the genotypes of *K. oxytoca* isolated from *Rhagoletis* could degrade pectic substances, a few of the genotypes showed only marginal activity.[488] However, other species of bacteria inhabiting the esophageal bulb could degrade pectic substances. Based upon the absence of pectinase activity in some of the genotypes of *K. oxytoca*, the authors were reluctant to assign pectin degradation as a functional role of the symbiotic bacteria. However, this should not preclude the possibility that some of the genotypes may possess other enzymes capable of degrading different plant matrix polysaccharides (e.g., hemicelluloses).

Investigation of bacteria associated with certain nonherbivorous dipterans has revealed that their symbiotic relationship with their host is somewhat similar to the symbiote-tephritid relationship, including the production of digestive enzymes, synthesis of various nutritional requisites, and/or the synthesis of volatile, metabolic by-products that act as attractants to adult flies. Bromel et al. isolated ten species of bacteria from larvae and wounds caused by the screwworm fly, *Cochliomyia hominivorax*.[65] They determined that these bacteria probably were producing nutrients, host attractants, and, in addition, bactericidal metabolites that prevented purulency of screwworm wounds. In an adjoining study, Gassner et al. determined that three of the isolates produced chitinases when grown in culture media.[204] One of the isolates, identified as *Pseudomonas fluorescens*, was embedded in the cuticle of the larvae. This observation led Gassner et al. to speculate that the production of chitinases is for the digestion of the cuticle during molting. Symbiote-produced chitinases are also found in cockroaches and termites for the digestion of fungi or for the retrieval of nitrogen from the exoskeleton (*vide supra*).[548] They further suggested that variability in host-selection and behavior by screwworm flies may be associated with different strains of symbiotic bacteria. Unlike the findings of Howard et al. with *Rhagoletis*,[259] Gassner et al.[204] found that electrophoretic isozyme patterns of symbiotic bacteria varies among screwworm flies according to host-selection behavior and locality. Bacterial contamination of growth media generally improves the growth of other flies as well.[197,218] However, no true obligate relationship was proven to occur between the bacteria and these flies. This is notable from the standpoint that unlike other insects, the gut pH of cyclorrhaphous flies is generally too acidic for the survival of bacteria.[182]

Lastly, it has been shown that the aquatic larvae of a detritivorous crane fly, *Tipula abdominalis*, acquires some enzymes capable of depolymerizing various plant matrix polysaccharides. These enzymes are provided by bacteria and protozoa inhabiting the lumen of the alimentary tract or midgut caeca.[206,371,506] Similarly, some of the carbohydrases (i.e., β-1,3-glucanases) occurring in the gut of blackfly larvae may also be produced by symbiotic bacteria.[370] However, the negligible cellulolytic activity found in both of these insects is supplemented by cellulases derived from fungi inhabiting the detritus ingested by these insects (*vide infra*).

d. Miscellaneous

Extracellular bacteria considered to be symbiotes have been observed or studied in a number of other insects. For the most part, the studies in this miscellaneous group of insects are less extensive than the more widely studied groups of insects reviewed above. From these miscellaneous studies, one of the more interesting roles invoked for microbial symbiotes is the production of enzymes for the digestion of wax. Surprisingly, aposymbiotic honey bees do not produce the carbohydrases able to degrade beeswax. These enzymes are presumably produced by symbiotic bacteria which may also provide various B-group vitamins.[103,236] Similarly, digestion of ester and fatty acid components of beeswax by the wax moth, *Galleria mellonella*, is performed by extracellular bacteria in the midgut.[152,208,433] The relationship of these bacteria to *Galleria* appears to be facultative in view of the fact that aposymbiotic *Galleria* can survive.[564] A number of other moth larvae are known to harbor symbiotes in their midgut. The most thoroughly studied of these bacteria-moth symbiotic relationships concerns the silkworm, *Bombyx mori*. Several studies have demonstrated the presence of aerobic bacteria and/or fungi in the midgut of the silkworm.[267,303,426,459] In general, the studies conclude that enteric microorganisms in the silkworm are ingested. Portier hypothesized that leaf-surface bacteria ingested by silkworm larvae contributed to digestion of cellulose.[459] Cellulase and proteolytic activity were also found to be produced by enteric bacteria and fungi of the silkworm by Jones et al.[295] The possibility that microorganisms provide essential nutrients (e.g., sterols and B vitamins) in the diet of the silkworm is suggested by the improved growth of silkworms on artificial diets that are contaminated with certain bacteria or fungi.[289,380] The failure of the sericulture industry initiated by colonists in Georgia may have been due to antimicrobial volatiles (terpenes) emanating from the bald cypress. Timber from this tree was used for the construction of the unviable silkworm-rearing facility.[294,295] In contrast, the microflora in the midgut of another moth, the Douglas fir tussock moth (*Orgyia pseudosugata)*, can metabolically convert terpenes to nontoxic compounds.[11] Bacterial symbiotes provide a similar service to scolytid beetles, which also attack coniferous trees (*vide supra*).[115] Bacterial symbiotes in two other species of moths, the turnip moth (*Scotia segetum*) and the almond moth (*Cauda cautella*), may play a role in maintaining dietary requirements and normal fecundity.[92,300] Except in the case of the almond moth, the type of symbiotic relation between bacteria and moth has never been established but is probably opportunistic.

The crop and ventriculus of both locusts and grasshoppers have an abundant microflora.[544] In a generalized survey of the known digestive enzymes in acridids, Morgan reported a number of enzymes that would appear to be of microbial origin, enzymes capable of degrading various plant matrix polysaccharides.[411] Low levels of cellulase activity occurred in 3 species of the 11 acridids surveyed. A variety of hemicellulases (i.e., laminarinase, lichenase, and xylanase) of moderate activity occurred in *Locusta migratoria*, and a wide range of β-glycosidases commonly appeared throughout the Acrididae. However, pectinase activity was not detected in any of the grasshoppers or locusts studied. No definitive attempts were made to determine if these enzymes were of microbial origin.

The observation of extracellular bacteria in the midgut and gastric caeca of various Heteroptera by Forbes[191] and later by Glasgow[213] are some of the earliest evidence of so-called symbiotic bacteria in insects. Ragsdale et al. cultured more than 30 species of bacteria and fungi from the southern green stinkbug, *Nezara viridula*.[475] Some of the bacteria isolated from the stomach proved to be potential plant pathogens and produced undefined phytotoxic enzymes. However, preliminary identification of the bacteria led to the conclusion that the stinkbugs acquired these bacteria from their environment. Bacteria and fungi inoculated into cotton during feeding by the cotton fleahopper, *Pseudatomoscelis seriatus*, induced synthesis of stress ethylene, causing early abscission of flowers, squares, and stunted growth of cotton plants.[163,227] These microorganisms were eventually isolated from the salivary glands of the fleahopper, leading to speculation that microbial enzymes in the saliva of the bug played a role in elicitation of ethylene synthesis by the plant.[375,378] Khan suspects that bacteria isolated from the third ventriculus of the cotton stainer, *Dysdercus fasciatus*, provide the bug with B vitamins that are absent from its host, cotton seeds.[301,302] In a different pyrrhocorid, a new gram-positive anaerobic bacteria was recently cultured from the midgut of *Pyrrhocoris apterus* and designated as a new symbiotic species of bacteria.[230]

Pectinases and, to a lesser extent, cellulase occur in many Hemiptera.[338,390,391] However, in many cases it is not known if these enzymes are supplied by microbial symbiotes. In the case of the induction of stress-ethylene synthesis by the cotton fleahopper (*vide supra*), it is now believed that pectinases in the saliva of this mirid produce elicitory oligosaccharides by degrading the cell wall.[376,377] Since these pectinases also occur in bugs whose salivary glands are devoid of microorganisms, it is speculated that these enzymes can also be produced by the insect.[376]

Most of the relationships between extracellular microorganisms and hemipterous insects may simply be casual.[522] The blood-feeding bugs (e.g., triatomines and cimicids)[22,23,558] and the plant bug, *Stenotus binotatus,*[98] which are known to possess intracellular microorganisms, are exceptions. A gram-negative, rod shaped bacterium in the hemolymph of the leafhopper, *Euscelidius variegatus,* may also be an exception.[472] This bacterium (designated as BEV) is transovarially transmitted. It appears to be a nonobligate symbiote in *E. variegatus*, but is pathogenic when injected into other species of leafhoppers.

B. EXTERNAL OR CUTICULAR SYMBIOTES

It has recently become apparent that insects have customarily established symbiotic relationships with microorganisms that reside outside the insect. In these cases, the symbiotes live either on the exoskeleton of the insect or somewhere within its biosphere. There are several comprehensive reviews on the biology and function of externally maintained symbiotes with their insect host.[25,194,234,362,363,365,520] In certain instances these microorganisms are "cultivated" by the host (e.g., attine ants). Like their internal counterparts, these external symbiotes provide the host with enzymes, nutrients, and metabolic assistance. However, these symbiote by-products are manufactured exogenously and, in many instances, their functional role is also performed outside the insect (e.g., digestion of cellulose prior to ingestion by the insect, breakdown of terpenoids to serve as pheromone precursors, etc.). In addition to supplying metabolic by-products secreted by or diffused from saprophytic microorganisms, certain insects have established an obligate symbiotic relationship wherein cultivated, symbiotic fungi serve as the chief if not sole source of food. In these cases, the insects have evolved digestive hydrolases and proteases (e.g., chitinases and β-1,3-glucanases) capable of depolymerizing the characteristic biopolymers of fungi. In these fungi-insect relationships, the fungi serve as a carbon and nitrogen source in addition to supplying vitamins and lipids (e.g., sterols). Such relationships have been shown to be well established in ambrosia beetles,[1,313,476] leaf-cutting ants,[200,241,467,486,549] termites,[26,491] beetles,[299,374] and wood wasps.[355] In fact, the strain of fungal symbiote which is established has been shown to affect the overall worker population size in colonies of attine ants[361] and fecundity in bark beetles.[526] Microorganisms in rotting fruit may also supply nutrients to fruit flies,[159,488] possibly the onion maggot,[495,554] or *Drosophila*.[304,519] A unique role of the symbiotic fungus of a gall-forming cecidomyid, *Asteromyia carbonifera*, is to act as a physical barrier to ovipositioning by torymid parasitoids.[550] Finally, one might consider the degradation of insecticides by certain soil-inhabiting microorganisms as an abstract or fortuitous form of ectosymbiosis.[381]

1. Acquisition of Digestive Enzymes

Most of the more recent studies on the acquisition of digestive enzymes from saprophytic or mutualistic ectosymbiotic bacteria or fungi have been by Martin and colleagues at the University of Michigan. They first demonstrated that a higher termite, *Macrotermes natalensis*, acquired active cellulase after ingesting its cultivated fungi.[366,367] Since this initial discovery, Martin and colleagues have found a number of additional insect-fungal/bacterial relationships wherein fungi or bacteria serve as an exogenous source of polysaccharases capable of digesting plant substances otherwise refractory to digestion by the insects. Many detritivorous aquatic insects possess enzymes that are capable of breaking down β-1,3-glucans, a linkage common to structural polysaccharides of algae and fungi.[370,371] It is not clear if these glucanases are of microbial origin. However, minor amounts of cellulase, hemicellulase, and/or pectinase activities have been detected in blackflies,[370] crane fly larvae,[371,506] caddis fly larvae,[373] and stone fly nymphs.[372] The acquisition of digestive enzymes from and/or utilization of fungi by mycophagous attine ants[50,368,369] and beetles[374] has also been assessed by Martin's group. In general, detritivorous and mycophagous insects possess different types of polysaccharases from those of herbivorous or xylophagous insects.[362,364,374] The former group generally possesses symbiote-supplied chitinase and laminarinase to degrade the structural biopolymers of fungi and algae, whereas the definitive acquisition of cellulolytic, pectinolytic, and/or hemicellulolytic enzymes from saprophytic fungi has been shown by Kukor and Martin to occur in wood wasps,[327] cerambycids,[328,330] and a wood louse.[329]

2. Biosynthesis of Semiochemicals

There are some general reviews on the potential role of symbiotic microorganisms in the synthesis of semiochemicals.[44,57,237,525,565] The most thoroughly studied cases of semiochemical production by ectosymbiotic microorganisms concern the Scolylitidae, but there are numerous other cases where microorganisms possibly synthesize insect trail pheromones, aggregating pheromones, host-locating kairomones, and possibly allomones. Kairomones, aggregation pheromones, or allomones are known to originate from debilitated plants that are infected with plant pathogens. Semiochemicals are frequently found in wounded plant tissues resulting from secondary infections by opportunistic or symbiotic microorganisms from infesting insects. The fungus *Gloeophyllum trabeum*, inhabiting rotten logs, synthesizes (*Z,Z,E*)-3,6,8-dodecatrien-1-ol, which is also a component of the trail pheromone of various subterranean termites.[260,382,383] A number of studies have implicated the various types of fungi and bacteria associated with bark beetles[194] in the biosynthesis of semiochemicals. Brand et al. showed that the mycangial fungi from *Dendroctonus frontalis* can oxidize *trans*-verbenol to the attack-density regulating pheromone verbenone.[54] It can also synthesize various bark beetle aggregation pheromones.[53,56] Similar observations were made by Borden et al. for *D. ponderosae*.[45] Moreover, free-living yeasts isolated from *Ips typographus* can interconvert *cis*- and *trans*-verbenol or convert *cis*-verbenol to verbenone.[344] Implication of bark beetle symbiotic microorganisms in pheromone production or establishment of infestations is represented in studies were the symbiotes are suppressed. Incorporation of streptomycin in the food of *I. paraconfusus* inhibited the production of the sex pheromones ipsenol and ipsdienol.[84] In a similar experiment, Chararas determined that microbial synthesis of *Ips* pheromones from monoterpenes does occur but that bacteria are not key to the synthesis of the pheromones.[104] Resistance of trees to bark beetle attack is associated with greater resin production and/or increase in monoterpenes.[292] Both of these factors have been shown to be toxic to the symbiotic fungi of *D. ponderosae* and *Scolytus ventralis*, respectively.[473,474]

A few studies show the involvement of bacteria or fungi in the biosynthesis of fermentation products or volatiles that act as host-locating cues. A braconid parasitoid of tephritid larvae uses fermentation products from rotting fruit to locate its host.[217] Microbial metabolic by-products of plant compounds also play a role in the production of kairomones for the seedcorn maggot,[167] ambrosia beetles,[408,482] the cabbage maggot,[178] the onion maggot,[251] and bark beetles.[214,455]

III. ENDOSYMBIOTES

Although determining the role of extracellular symbiotes in insects has been relatively successful, understanding the role of endosymbiotes remains elusive. However, with the increased application of molecular biological techniques in the study of insect endosymbiotes, reliable information is increasingly available. Using light microscopy, some of these endosymbiotes were originally described as yeasts.[345,528] They have since been described as plasmids,[342] rickettsiae,[334] Blochmann bodies,[337] or a variety of bacteria,[79,414] and their true taxonomic identities are still in question.

The taxonomic and biochemical characterization of insect endosymbiotes will probably soon be routinely accomplished using recombinant DNA techniques, and it is worthwhile to acquire an appreciation of these new techniques. In order to achieve this appreciation, one must also understand much of the research that has occurred on insect endosymbiotes in the past. Because of the inherent difficulties in studying an intracellular, obligate symbiote, some of the conclusions on the taxonomy and biochemistry of insect endosymbiotes made in the past have since been found to be incorrect or are uncertain. Nevertheless, much of the previous descriptive work on the ultrastructure, histology, and morphology of insect endosymbiotes remains valid. Hence, an appraisal of past and present research on insect endosymbiotes can provide a perspective of how the field has advanced to the point at which it now stands. In order to achieve this perspective, a review of earlier research on endosymbiotes is presented prior to a review of the more recent literature on the biochemistry, metabolism, and molecular biology of insect endosymbiotes. Comprehensive reviews of earlier research on insect endosymbiosis are available.[67,76,79,130,240,311,312,414,485,521,522,524]

A. HISTOLOGY, ULTRASTRUCTURE, AND HOST NUTRITION

Endosymbiotes of insects are generally housed within specialized cells of insects commonly referred to as mycetocytes. In some insects, individual mycetocytes may be unconnected to any other mycetocyte, internal tissues, or organ of the host insect. Frequently, however, they are clustered together (often more than 100) to form an organ-like structure, originally described in aphids, leafhoppers, and weevils, termed a mycetome.[542] Mycetocytes may also occur within the cells of tissues of various insect organs (e.g., malpighian tubules and salivary glands).[79,414] Koch originally noted that insects having a diet restricted to a very narrow range of host tissue invariably possess intracellular microorganisms (viz., endosymbiotes).[310] He further postulated that these symbiotes provide some form of nutritional requisite that is missing or in low supply in the natural (and presumably inadequate) diet of the insect host. Furthermore, he noted that the importance of the endosymbiotes in host survival is reflected in the fact that a mechanism for their transmission from adult host to offspring has evolved.

Endosymbiotes are found in numerous insects and arachnids which are not herbivorous. There are a number of studies which describe the role of endosymbiotes in hematophagous arthropods such as bedbugs,[95,102] lice,[12-14,165,166,454,471,481] tsetse flies,[379,439-441,480,510] mosquitoes,[185,389,567] ticks,[8,80] and triatomine bugs,[22,58,205,215,242,472,559] with reviews available.[23,96] These studies on hematophagous insects are not discussed in detail except where they relate to understanding the role of endosymbiotes in insects in general. It is assumed, however, that the symbiotes provide nutrients to the host and, for the most part, their biology is similar to the endosymbiotes of herbivorous insects.

The two most thoroughly studied groups of herbivorous insects which possess endosymbiotes are the Homoptera and Coleoptera. As in nonherbivorous insects, the presence of endosymbiotes in certain members of these herbivorous insects generally connotes a relatively restricted diet confined to one or two particular tissue-types of the host plant. However, many cockroaches, which are less restrictive in their diet, also possess endosymbiotes. Because of the vast amount of literature available, endosymbioses in these three groups of insects are reviewed individually.

1. Homoptera

The most recent comprehensive coverage of the literature on endosymbiotes of the Homoptera was published by Houk and Griffiths in 1980.[254] It is noteworthy that in their introductory remarks they lamented that little progress beyond the descriptive stage had been made in the field of insect endosymbiosis since the work by Brooks on cockroach endosymbiotes some 25 years earlier.[66] Today, we are on the verge of some long-awaited breakthroughs in this field. Manipulation of host-symbiote relationships with antibiotics was the chief method used to investigate endosymbiosis in insects, but with the increased use of recombinant DNA techniques, more direct and reliable information is increasingly becoming available. Nonetheless, much of the early descriptive work on the biology of insect endosymbiotes is of value.

Significant additions to information on the descriptive taxonomy and morphology of insect endosymbiotes are few since the review of Houk and Griffiths. Within the Homoptera, descriptive studies of the morphology and ultrastructure of intracellular microorganisms (viz., symbiotes) have been performed on leafhoppers,[99-101,233,318-323,349-351,395,446,490,499-504] planthoppers,[332,333,429,434] psyllids,[97] aphids,[78,220,243,245,246,266,334,336,337,339,388,458,542,545] and scale insects, including mealybugs.[77,93,94,186,346-348,537]

When only light microscopy was available, there was debate as to whether endosymbiotes housed within mycetocytes were indeed microorganisms, a form of cell organelle, or an artifact.[336,485,535] However, the early electron micrographs of the so-called endosymbiotes in the oocytes of cockroaches[82,219] and grain weevils,[423] and eventually the observation of nonnuclear strands of DNA, offered strong evidence that the endosymbiotes were of microbial origin. Shortly following these studies, electron microscopic studies of homopteran endosymbiotes were initiated and revealed that the mycetocytes of aphids and leafhoppers contained bacteria-like microorganisms whose typical rod shape could vary between species and type of endosymbiote. Some species of aphids (e.g., pea aphids) possess two types of endosymbiote: a primary coccoid endosymbiote and a secondary bacilliform symbiote. The two types of symbiote are located in separate mycetomes and occur at a ratio of $\approx$100 to 1 (primary to secondary).[254] Attempts to classify endosymbiotes have been based on histochemistry, DNA guanine + cytosine content, and/or morphology. The prokaryotic endosymbiotes of aphids are generally gram-negative eubacteria[254] and have been described as rickettsias,[334] L-form bacteria,[246,555] and *Rickettsia, Chlamydia*, or mycoplasma-like.[257,388] None of these designations appears to be valid based upon recent evidence of the nucleotide sequence of endosymbiote 16S ribosomal RNA (*vide infra*). Leafhoppers also generally possess two (designated types "a-" and "t-") and in some species three types of prokaryotic endosymbiotes and a eukaryotic, yeast-like endosymbiote housed within mycetocytes that are clumped together as an abdominal mycetome.[99,320,321,350,351,393,502] Planthoppers do not possess a mycetome. Their yeast-like endosymbiotes are housed in mycetocytes dispersed throughout the fat body and, in some species, rickettsia-like prokaryotic symbiotes are found either extracellularly or in mycetocytes.[434] Homopteran mycetocytes possess a plasma membrane of microbial origin and another membrane, surrounding the endosymbiote, of host origin.[233,266,320,349,545] In studies of the ultrastructure of endosymbiotes of aphids, it was determined that the endosymbiote possessed two discernible membranes, an inner plasma membrane and an outer cell wall of peptidoglycan (i.e., *N*-acetylglucosamine with diaminopimelic and *N*-acetylmuramic acids) typical of gram-negative bacteria.[222,243,258] However, one of the major contributions of these ultrastructural studies was the observation of electron-dense "particles" or "vesicles" associated with the peripheral cytoplasm or plasma membrane of the symbiote and/or the cytoplasm of the mycetocyte (host).[321,350,351] The observation of such structures was perceived to be visible evidence of an active mechanism for the physical exchange of biosynthetic products between symbiote and host cell including nutritional requisites for the host.[101,222]

The idea that homopteran endosymbiotes supplied nutrients to the host arose from studies involving artificial diets.[15,18,324,404,406] Artificial diets were eventually formulated for a number of aphids, including *Myzus persicae*,[138,140-143,400,401,405] *Neomyzus circumflexus*,[17,172] *Acyrthosiphon pisum*,[2,3,16,19,20,360,483,484,512,513] *Aphis fabae*,[139,168,341] *A. gossypii*,[17,540,541] and *Schizaphis graminum*,[161] as well as planthoppers and leafhoppers (*vide infra*) in order to study nutritional requisites supplied by endosymbiotes.

Direct culture studies of extracellular symbiotes of insects had established that symbiotes provide nutrients to the host. Studies of endosymbiotes, however, did not have the benefit of *in vitro* culturing and speculations on their metabolism have been drawn using indirect methods of analysis.

The primary experimental method for studying endosymbiote metabolism has involved the treatment of the host insect with antibiotics,[174,175,221,291,402,403,498,514] lysozyme,[169] or heat[186,436] in an attempt to diminish or eliminate the endosymbiotes. In a more refined approach, Ishikawa and colleagues used an array of antibiotics on the pea aphid. These antibiotics were eukaryotic (host cell) or prokaryotic (endosymbiote) specific and disrupted transcription or translation during protein synthesis by the cell.[283] In this fashion they attempted to isolate the products of gene expression of the host cell vs. its symbiotes. The results of their studies are discussed in detail in the next section. In any case, aposymbiotic insects generally show a significant reduction in their rate of development and fecundity, and their progeny are usually sterile.[254] These observations, plus the evolution of a fairly elaborate method for the transovarial transmission of endosymbiotes,[79,245] indicate that endosymbiotes play a significant biological role in the survival of sap-feeding insects. Attempts at *in vitro* culturing of endosymbiotes of aphids and leafhoppers to directly observe their biosynthetic capabilities have generally resulted in only short-term success.[244,253,255,269,270,331-333,394,446,500,505] Characterizations based on these cultures are of questionable reliability.

a. Amino Acids

By using artificial diets and in some instances *in vivo* cultures, a number of conclusions could be reached regarding biosynthesis of amino acids, vitamins, and lipids by endosymbiotes. Aphids are incapable of fixing nitrogen,[509] and some investigators believe that the endosymbiotes of sap-feeding insects are capable of supplying their hosts with certain essential amino acids. After the composition of amino acids in plant sap and honeydew was compared, it was concluded that tryptophan is synthesized by the symbiotes of *Tuberlachnus salignus*,[398,399] *Megoura viciae*, and *Dactynotus henrici*,[148] but not those of *A. gossypii*.[541] There are varying assessments of the ability of aphid symbiotes to synthesize the sulfur-containing amino acids methionine and cysteine. Cysteine can be formed through transulfuration of serine by homocysteine (which is produced by the demethylation of methionine) in animals. However, prokaryotes can directly synthesize cysteine through the sulfhydrylation of serine and synthesize methionine by the transulfuration of homoserine by cysteine.[518] Ehrhardt demonstrated that aposymbiotic aphids of *N. circumflexus* containing symbiotes utilized sulfer to form methionine or cysteine, whereas aposymbiotic aphids did not.[170,172] Leckstein showed that *M. persicae* was capable of performing transulfuration of these amino acids, but that it ws not dependent on the symbiotes.[341] Nonetheless, both methionine and/or cysteine were essential amino acids in the artificial diets of *A. pisum*,[3,483] *M. persicae*,[140,402] and *A. gossypii*.[540] Cysteine, as a sole amino acid in artificial diets, did not support growth of the pea aphid.[515]

Whether aphid endosymbiotes supply the other amino acids to their hosts is even more puzzling. Separate studies have come to differing conclusions as to the essential amino acids required by the pea aphid. In one study, the artificial diet of the pea aphid required all ten essential amino acids (i.e., arginine, histidine, isoleucine, leucine, lysine, methionine, phenylalanine, threonine, tryptophan, and valine) in addition to cysteine to support growth.[484] Other studies have shown that the pea aphid can grow without the full complement of essential amino acids.[360,483,515] The artificial diet of Dadd and Krieger for *M. persicae* contained only histidine, isoleucine, and methionine.[140] A recent study by Srivastava and colleagues indicated that different biotypes of the pea aphid have different requirements for amino acids.[516] They also observed that single, nonessential amino acids were able to support growth in the pea aphid[515] and that protein synthesis occurred when only [^{14}C] leucine was supplied to the aphid.[513] Aposymbiotic progeny from green peach aphids treated with various antibiotics could not grow unless all ten essential amino acids were included in their diet.[402]

The above observations have led to the hypothesis that endosymbiotes of aphids are capable of supplying their hosts with various essential amino acids that may be deficient or missing in their food source.[402,516,517] However, aphids excrete large amounts of amino acids in their honeydew. This may indicate that the supply of amino acids in the phloem-sap of plants is in excess of that required by aphids.[511] Nonetheless, during senescence the supply of amino acids is increased due to proteolysis in the plant and aphid performance usually improves.[179-181,553] The availability of amino acids in the host plant for sap-feeding insects may not be any lower than for insects which do not maintain an obligate endosymbiotic relationship with microorganisms. Thus, whether there is a requirement for supplementation of amino acids by endosymbiotes in the natural diet of aphids is uncertain.

b. Sterols

Another biosynthetic role hypothesized for homopteran endosymbiotes is the biosynthesis of lipids, especially sterols. Insects are incapable of the *de novo* synthesis of sterols.[119] They are unable to condense farnesyl pyrophosphate to squalene and lack squalene epoxide cyclase, an enzyme which catalyzes the cyclization of squalene to lanosterol.[158] In view of these biochemical inadequacies, phytophagous insects must procure sterol precursors (e.g., the C_{28} and C_{29} phytosterols) from their host plant.[529] These sterol precursors can then be dealkylated to cholesterol or other utilizable C_{27} sterols. Once dealkylated, they can be used in the construction of cellular membranes and/or converted to the insect molting hormones, the ecdysteroids.[530]

Aphids apparently do not require an exogenous source of sterol due to the biochemical capability of their endosymbiotes to make sterols.[435] However, the arguments for this conclusion have their basis in indirect dietary studies and lack direct biochemical evidence. The idea that aphids may not be able to acquire all their nutritional requisites from phloem sap, including sterols, was asserted in 1968 and earlier.[170,311] Some studies showed that aphids could be maintained on artificial diets that lacked sterols.[2,140,142,171,512] There is a more recent observation that aphids of *M. persicae*, made aposymbiotic by treatment with chlortetracylcine, can produce viable offspring only when reared on a sterol-supplemented diet.[156] The fact that symbiotic *M. persicae* develop normally on sterol-free diets implies that there is *de novo* biosynthesis of sterol by the symbiotes. It was also observed that aphids rendered aposymbiotic with antibiotics could not convert [^{14}C]acetate to sterol while aphids with symbiotes could.[173] Moreover, isolated endosymbiotes from the pea aphid synthesized cholesterol when incubated in [^{14}C] acetate or [^{3}H] mevalonic acid.[256] Histochemical analyses of the plasma membrane of pea aphid endosymbiotes also detected sterols.[223,224,252]

However, other factors make the role of aphid endosymbiotes in sterol biosynthesis debatable. For example, phytosterols have been identified in the honeydews of a variety of sap-feeding insects, suggesting that sap-feeding insects can procure sterols from their host plants.[192,438,527]

Recent metabolic evidence indicates that aphids do not synthesize sterols and therefore must acquire phytosterols from their host plants. Greenbugs, *Schizaphis graminum,* fed [^{14}C] mevalonic acid could not synthesize any radiolabeled sterols nor intermediates in sterol synthesis (e.g., farnesol, squalene, 2,3-oxidosqualene or 4,4-dimethyl, and 4-monomethyl sterols). The aphids did, however, acquire C_{28} and C_{29} phytosterols from their host plant. The aphids were then able to dealkylate these phytosterols and convert them to cholesterol.[87] The sterols that were detected in the earlier studies[173,223,224,252,256] (*vide supra*) were probably a series of C_{24} to C_{30} long-chain fatty alcohols and their esters. These compounds show thin-layer and gas chromatographic behavior (identification techniques used in the earlier studies) similar to sterols. Furthermore, aphids are able to convert mevalonic acid to long-chain fatty alcohols via the *trans*-methylglutaconate shunt.[431]

How aphids are able to develop on sterol-free artificial diets is still a mystery. In general, aphids sequester only very small quantities of sterols ($<$18 ng per aphid).[87] Perhaps the so-called sterol-free diets of earlier studies were contaminated with minute amounts of sterol. These contaminants could come either from impurities in the holidic diets (e.g., within the waxes of Parafilm® or Nescofilm®, which are standardly used to overlay or secure aphid diets) or from contaminating microorganisms,[431] although the latter has recently been shown to be unlikely.[156]

Lastly, the ultrastructure, histochemistry, and morphology of both primary and secondary symbiotes of a number of aphids, including the pea aphid, indicate that they are eubacteria.[254] Based upon the nucleotide sequences of the rRNAs of both symbiotes of the pea aphid and a number of other species of aphids, it appears that aphid symbiotes are taxonomically related and belong to the gamma subdivision of the purple bacteria (*vide infra*).[543] Because bacteria in this group are incapable of sterol biosynthesis,[432] it would be truly extraordinary if aphid endosymbiotes synthesized sterols.

Some leafhoppers and planthoppers have been reared on artificial diets devoid of sterols.[316,317,396,397] Nevertheless, other leafhoppers presumably need a supply of sterol in their food.[250,315] It is suggested that the eukaryotic, yeast-like endosymbiotes synthesize and provide sterols to those leafhoppers and planthoppers that can be sustained on sterol-free diets.[435] Various rice planthoppers have been shown to contain cholesterol, two of the more common phytosterols (i.e., β-sitosterol and campesterol), and 24-methylenecholesterol, an intermediate in the conversion of campesterol to cholesterol.[529] Incubation of 3-d-old nymphs of smaller brown planthoppers, *Laodelphax striatellus,* at 35°C greatly reduces the number of their yeast-like symbiotes.[436] Other results of the heat treatment are reduction in the amount of 24-methylenecholesterol in the insect and disruption of normal ecdysis to the adult stage. Supplementation of cholesterol in the artificial diet partially reverses the effects of the heat treatment. These observations led to the conclusion that the symbiotes are capable of synthesizing 24-methylenecholesterol and that this sterol is the origin of ecdysone in the insects.[437] Moreover, the lack of 24-methylenecholesterol in another planthopper, *Nilaparvata lugens,* and the requirement of sterol in its diet was associated with the fact that this planthopper only possesses prokaryotic symbiotes.[435] It was recently reported that yeast-like symbiotes isolated from the eggs of *L. striatellus* synthesized and exuded ergosterol into the culture medium.[437] Ergosterol can be a source of sterol for a wide range of insects.[486] However, further, unambiguous biochemical scrutiny will be required to substantiate the biosynthesis of sterols by symbiotes of leafhoppers or planthoppers. It also remains to be determined if such a biosynthetic capability is of any physiological and/or ecological consequence since these insects can obtain sterols from the host plant.

c. Pigments, Antibiotics, Vitamins, and Carboxylesterases

Other roles suggested for the endosymbiotes of homopterans are the synthesis of aphid pigments, antibiotics, and vitamins.

The possibility that symbiotes may provide antibiotics for protection against infection by pathogenic microorganisms is based on *in vitro* culture of planthopper endosymbiotes.[195] Two flavanoids, 2,4-diacetylphloroglucinol and pyoluteolin, were identified in the culture medium of symbiotes from the planthopper *Sogatella furcifera.* An antibiotic, andrimid, was isolated from the culture broth of symbiote strain N1-T-C1-1W from *N. lugens.*

Another putative role for homopteran symbiotes is the provision of vitamins to the host insect. The aphid *Neomyzus circumflexus* was reared for ten generations on an artificial diet devoid of the B-complex vitamins: B_2, pantothenic acid, and biotin.[172] The other B-complex vitamins in addition to ascorbic acid (vitamin C) and the cyclitol, *meso*-inositol, were required, however. The artificial diet of the green peach aphid requires all nine of the B-complex vitamins,[143] while that for the pea aphid contains all B-complex vitamins aside from vitamin B_{12}.[2] Whether there is provision of certain vitamins to the host insect by its symbiotes needs to be biochemically demonstrated.

There is speculation that intracellular symbiotes contribute to resistance of aphids against organophosphorous pesticides.[7,10] Amiressami and colleagues noted that the number and size of endosymbiotes harbored in ethylparathion-resistant green peach aphids and Demeton-*S*-methyl-resistant hop aphids differed from those in susceptible biotypes.[7,9] Also, the degree of vacuolization in the cytoplasm of the mycetocytes of the resistant biotypes was greater than in susceptible biotypes. In light of these observations, they concluded that the endosymbiotes of the resistant biotypes were physiologically more active and were playing some role in providing resistance against the pesticides. However, Devonshire and colleagues have recently shown that carboxylesterases in resistant biotypes of *M. persicae* are derived from host mRNA (i.e., poly[A] + RNA).[151] Furthermore, the genetic basis of resistance is related to the karyotype of resistant clones and greater esterase production is through gene amplification in the host DNA.

The evidence that endosymbiotes assemble the various pigments found in aphids, although indirect, is fairly convincing. Brown, at the Centro de Pesquisas de Produtos Naturais, and colleagues at the National Institutes of Health in Bethesda, MD determined that the yellow "aposematic" pigmentation of the oleander aphid, *Aphis nerii,* is a composite of four polyketide glucosides and their derivatives.[73-75] These pigments were isolated from two different populations of aphids. One cohort was collected in Tucson, AZ from oleander, *Nerium oleander,* and the other population was collected from milkweed plants, *Asclepia curassavica,* in Rio de Janeiro, Brazil. The fact that these polyketide pigments appear in populations of aphids that differ both geographically and ecologically (tropical vs. dry; milkweed vs. oleander) indicates that the pigments are natural products of the aphids.[73] In addition to the polyketides, an unusual triglyceride, 2-*trans*, *trans*-sorbo-1,3-dipalmitin, was also isolated from the aphids. The presence of these unusual natural products in an insect, in addition to their resemblance to polyacetate fungal metabolites, suggests that their biosynthesis is by microorganisms, namely the endosymbiotes. Moreover, pigmentation differences between biotypes of aphids may also be associated with metabolic differences between their symbiotes.[72]

2. Coleoptera

The major focus on endosymbiosis in beetles is with beetles that are pests of storage products. In most early studies, elimination of symbiotes by antibiotics, heat treatment, or fumigants indicated that symbiotes provided amino acids, vitamins, or other growth factors that were restricted in the austere diets of these insects.[447] The insects studied were the cigarette beetle (*Lasioderma serricorne*),[41,296-298,392,393,448,450] the drugstore beetle (*Stegobium paniceum*),[41,188,306,448,450] the furniture beetle (*Anobium punctatum*),[31] the powder post beetle (*Lyctus linearis*),[307] the sawtoothed grain beetle (*Oryzaephilus surnamensis*),[308,309,314,449] several bostrichids,[357,358] and the confused flour beetle (*Tribolium confusum*).[288]

Within the Coleoptera, the Curculionidae have the most extensive relationship with endosymbiotes.[79] Of these, the most thoroughly studied have been the pleomorphic gram-negative bacteria found in the rhynchophorid grain weevils (species of *Sitophilus* = *Calandra*), *S. oryzae* (= *S. sasakii*), *S. granarius*, and *S. zeamais*.[79,414] These endosymbiotes occur throughout all life-stages of the weevils.[424] The symbiotes are located in a wide variety of tissues and vary greatly in shape and size. These morphological characteristics vary depending on the tissue in which the symbiotes are located and/or the species or strain of *Sitophilus*.[420] The shape and size of these symbiotes can be used to distinguish the sibling species of the small rice weevil, *S. oryzae*,[416] and the maize weevil (also named the large rice weevil and the corn weevil), *S. zeamais*.[423] In both *S. oryzae* and *S. zeamais*, the endosymbiotes can be found in a tubular, horseshoe-shaped mycetome that enfolds the ventral portion of the foregut, in the ovaries, or in the gastric caeca.[356,358,415,418,427,428] In *S. granarius* the symbiotes are pleomorphic and vary greatly in size (from 3 to 25 μm). They are located within the same tissues as those observed for *S. oryzae* and *S. zeamais*, with the exception of a small form that can be found in the prostate and male accessory glands and in the malpighian tubules of both sexes.[34,225,226,354,493,494] Differences between the light- and dark-colored strains of *S. granarius* are associated with differences in their endosymbiotes. The light-colored strain appears not to possess the large symbiote normally located in the foregut mycetome and midgut caeca.[358,359,419,420,494] Elimination of the larger symbiotes in the dark-colored strain of *S. granarius* by treating of weevils at 31°C[422,494] or with methylbromide[425] results in offspring that resemble the aposymbiotic strain, although these offspring are paler and smaller. Generating aposymbiotic weevils of *S. oryzae* is difficult due to the ability of their symbiotes to withstand heat treatment.[422]

Attempts to culture *in vitro* the endosymbiotes of *Sitophilus* have generally resulted in the production of contaminants and have been basically unsuccessful.[131,144,417] Hence, it is difficult to accurately describe the biochemistry and taxonomy of *Sitophilus* endosymbiotes. Affirmation of this true prokaryotic nature came with the detection of DNA in the large symbiote of *S. granarius*.[421,508] Based on morphological characters, Bhatnagar and Musgrave assigned the large symbiote of *S. granarius* to the Myxobacterales.[34] An elegant experiment to determine the taxonomic nature and evolutionary relationship of *Sitophilus* endosymbiotes to endosymbiotes of other insects was performed by Dasch.[145] Using CsCl gradient ultracentrifugation, he determined the base compositions (i.e., percent guanine and cytosine [% G + C]) of endosymbiote DNA of the three species of *Sitophilus*. Differences in base composition paralleled differences in the morphology of the endosymbiotes associated with each species: *S. granarius* (50% G + C), *S. oryzae* (55% G + C), and *S. zeamais* (50 and 54.5% G + C). The latter finding of two different DNAs from the symbiote colony of *S. zeamais* indicates that this species harbors a second, previously undetected symbiote. These observations led to the conclusion that the endosymbiotes in all three species of *Sitophilus* are different. Based on their DNA base compositions, morphology, and histochemistry, Dasch speculated that they may be related to the *Saprospira* of the Flexibacteria. Moreover, he suggested that they might be placed in a new genus in the Cytophagaceae. In any case, the endosymbiotes of the *Sitophilus* weevils are almost definitely not in the Myxobacterales as previously suggested.[34] A recent suggestion that the endosymbiote of *S. granarius* is *Staphylococcus albus*,[235] as determined by cultural studies, requires further verification.

The symbiotes in *Sitophilus* weevils have also been difficult to investigate because of the difficulty to culture them *in vitro*. However, unlike with the Homoptera where antibiotics or other treatments are required to achieve aposymbiosis, there are *Sitophilus* weevil strains that are inherently aposymbiotic. A comparison of aposymbiotic and symbiotic strains of *S. oryzae* indicated that the symbiotes provide their host with riboflavin, pantothenic acid, and biotin.[556] Endosymbiotes also appear to be involved with the metabolism of aromatic amino acids[555] and in determining the composition of amino acids in peptides and proteins.[557] Recently, Gasnier-Fauchet and colleagues performed extensive experimentation showing the role of symbiotes in the metabolism of methionine in *Sitophilus* weevils. In a survey of the amino acid levels, they discovered higher levels of sarcosine, a methionine-derived amino acid uncommon in insects, and lower levels of methionine sulfoxide in aposymbiotic strains of *S. oryzae* as compared with a symbiotic strain. The synthesis of sarcosine in the aposymbiotic larvae of *S. oryzae* was tentatively shown to be due to the presence of an *N*-methyltransferase that was inhibited in symbiotic larvae.[203] Higher levels of methionine sulfoxide in the symbiotic strain were the result of oxidation of the sulfur atom inside the mycetocyte.[201] They found the same relationship of sarcosine and methionine sulfoxide in aposymbiotic and symbiotic strains of *S. zeamais*.[202] Thus, although the symbiotes of these two species of weevils are different, they appear to have homologous roles in the metabolism of methionine. When aposymbiotic and symbiotic strains of weevils were reared on three grain cereals having different contents of methionine, there was no correlation between methionine content of the cereal and levels of sarcosine or methionine sulfoxide in the insects.[202] After consideration of these observations, it is uncertain that the metabolism of methionine by weevil symbiotes provides a significant contribution to weevil survival.

One can only speculate as to the benefits these symbiotes provide *Sitophilus* weevils. It may be possible that their most significant biological contribution to the host has been overlooked. Weevils of both *S. oryzae* and *S. zeamais* have high levels of pectinase and hemicellulase activities (>100 nkatals) concentrated in the head and foregut. In addition, they possess moderate levels (≈10 nkatals) of carboxymethylcellulase and a variety of β-glycosidase activities.[89] It is tentatively worthwhile to believe that these enzymes are biosynthesized by the symbiotes. If this is the case, the endosymbiotes would enable the weevils to utilize the polysaccharides of cereal grains as a carbon and/or energy source more efficiently. Alternatively, these polysaccharases may simply facilitate the penetration of the seed coat of the grain.

3. Cockroaches

Most of the interest in the intracellular symbiotes of cockroaches is inspired by the potentially 300-million-year-old relationship between cockroaches and their symbiotes.[79] The cockroaches could perhaps serve as a model where evolutionary divergence between cockroach symbiotes parallels speciation in the host insects. Also of fundamental interest is the question of similarity between symbiotes in cockroaches and termites, since the cockroach *Cryptocercus* and the termite *Mastotermes* shared a common ancestor, *Pycnoblattina*, over 200 million years ago.[125] With respect to their biological role, some research has tentatively indicated that the endosymbiotes of cockroaches provide amino acids to the host and are involved in the metabolism of urates.

A number of early studies described intracellular bacteroids in the fat body and oocytes of a variety of cockroaches.[69,70,82,207,211,212,219] To gain more understanding on the taxonomy and biology of these symbiotes, attempts were made to culture them *in vitro*.[71,211] These cultures led to the description of numerous tentative identifications of the symbiotes. However, Brooks later surveyed the symbiotes in all cockroaches and found that their tissue location and morphology were astoundingly similar.[68] He discounted the identifications of earlier studies and classified all cockroach endosymbiotes as one species, *Blattabacterium cuenoti*. Dasch later surveyed the DNA base compositions of endosymbiotes from the fat body portion of ten species of cockroaches.[145] Surprisingly, he found the % G + C composition of the symbiote from nine of the ten species to be similar (26 to 28%), the exception possibly due to error because of the small amount of DNA available. Dasch's results concur with Brook's decision to classify all cockroach symbiotes as one species. This lack of morphological and genetic radiation in the endosymbiotes of modern cockroaches may indicate that these symbiotes were acquired only recently. Moreover, the relatively low % G + C composition of the symbiotes may reflect a low genetic code limit similar to that of mitochondrial and chloroplast DNAs.[24,487]

Only a few studies have focused on the potential biological role of cockroach endosymbiotes. The endosymbiotes do play a role in the storage excretion of uric acid from the fat body of cockroaches.[155] Ultrastructural studies show that the urate cells, which contain deposits of uric acid, are located adjacent to the mycetocytes within the fat body.[146] Uric acid produced in the trophocytes is transported to the urate cells by intra- and intercellular channels. Histochemical studies show that the symbiotic bacteria are able to metabolize the uric acid, thus purging the insect of the urate deposits.[560] Another minor role for the symbiotes may be in the synthesis of sulfur-containing amino acids. Cockroach endosymbiotes are able to metabolize inorganic sulfur.[239] However, this ability may just be fortuitous and it is not considered a major role.

B. MOLECULAR BIOLOGY OF APHID ENDOSYMBIOTES

There is currently a great deal of confusion surrounding the biological and taxonomic identity of aphid symbiotes. Prior to 1980, much of the research on aphid endosymbiotes was descriptive. Attempts to define their taxonomy were chiefly based on ultrastructural and histochemical studies. Other efforts to elucidate their biochemical role had to rely on indirect means of observation (e.g., use of antibiotics in diets, etc.) as a result of the inability to grow aphid endosymbiotes outside of their host. However, recombinant DNA techniques have recently been used to overcome many of the obstacles that curtailed the classification of aphid endosymbiotes in the past. The nucleotide sequences of the genes encoding for the RNAs of both endosymbiotes of the pea aphid have been used to approximate the bacterial ancestry of these symbiotes.[543] Now that this technique is in use, the identity and relatedness of the endosymbiotes in other species of aphids and insects may soon be determined. Moreover, use of genetic engineering techniques should illuminate the biochemistry and thus the biological role of these endosymbiotes.

1. Nucleotide Sequencing Studies of Endosymbiote rRNA

The sequence of nucleotide bases in genes encoding for rRNAs (sometimes referred to as rDNAs) is stable over evolutionary time relative to sequences of other genes.[561] This is perhaps because ribosomes are an integral component of the protein-synthesizing apparatus of all organisms. Because of this universality of function, the degree of sequence alignment in rRNAs (or rDNAs) between taxa is a useful measure of their phylogenetic relatedness.[445] Moreover, there are certain regions within rRNA where the nucleotide sequences are unique to a given taxon. These regions are referred to as signature sequences. In bacteria there are three rRNAs, the 5S, 16S, and 23S having ≈120, 1600, and 3000 nucleotides, respectively. For phylogenetic studies, the nucleotide sequencing of the 16S rRNA is generally preferred because of its relatively higher content of information (compared to the 5S rRNA). By comparing sequences of oligonucleotides of 16S rRNAs, Woese and colleagues were able to outline the genealogy of approximately 400 species of bacteria.[193,562,563] More recently, determining entire nucleotide sequences of 16S rRNAs and their sequence alignments has become a fairly workable proposition. These sequencing methods have been facilitated by dideoxynucleotide sequencing protocols[492] and rRNA-specific oligodeoxynucleotide primers that facilitate transcription and eventual cloning of complementary portions of rDNA.[335] These procedures result in more genetic information for rRNA sequence alignments and construction of evolutionary dendrograms.[444]

In the laboratory of Baumann at the Department of Bacteriology, University of California, Davis, Unterman has recently cloned the rRNA genes of both symbiotes of the pea aphid.[543] The protocols used to achieve this were made more difficult by the fact that the DNAs of both the host and the symbiote had similar base compositions (27% G + C).[282] Hence, their DNAs could not be separated by CsCl equilibrium gradient centrifugation. The total aphid DNA was digested with restriction enzyme(s) and the resulting restriction fragments were separated by gel electrophoresis. The rDNA restriction fragments were hybridized with a ^{35}S-adenosyl-labeled DNA probe of the *Escherichia coli* rRNA gene (i.e., Southern hybridization). The hybridized fragments were then detected by autoradiography. Once the symbiote restriction fragments were located on the gel, that portion of the gel was excised and the contained DNA electroeluted. After electroelution the fragments were cloned into *E. coli.*

The results of their research will eventually lead to the accurate assignment of the taxonomy of symbiotes of aphids. Analysis based on 603 nucleotides of the 16S rRNA between the two endosymbiotes shows an 82% similarity. The similarity of the sequences of the primary and secondary endosymbiotes to the rRNAs of other bacteria were as follows: *E. coli* (84, 93%), *Proteus vulgaris* (83, 90%), *Pseudomonas testosteroni* (67, 70%), *Rochalimaea quintana* (a rickettsia) (63, 65%), *Bacillus subtilis* (60, 63%), and *Mycoplasma capricolum* (55, 58%). Hence, the degree of similarity between the two endosymbiotes is no greater than their similarity to the enteric bacteria, *E. coli* and *P. vulgaris*. Moreover, neither endosymbiote is closely related to the rickettsia, *Pseudomonas,* or *Mycoplasma*. This is especially noteworthy since on the basis of morphology, histochemistry,[254] and/or nucleotide base composition,[282] previous researchers proposed that the endosymbiotes of the pea aphid may belong to one of these groups of bacteria. However, signature sequences within the rRNA of the endosymbiotes suggest that both endosymbiotes are phylogenetically related within the gamma subdivision of the purple bacteria. Of even more interest is the discovery that both endosymbiotes possess only one rRNA operon. By analogy with other bacteria, this finding indicates that the doubling time (growth rate) of the endosymbiotes is low.

2. Protein Synthesis and Gene Expression in Aphid Endosymbiotes

The possibility that genetic expression in pea aphid endosymbiotes is limited has also been suggested by Ishikawa. However, the hypothesis proposed on protein synthesis in aphid endosymbiotes is somewhat novel.[281] Research has led him to conclude that the endosymbiotes synthesize only one protein *in vivo*. The gene encoding this protein resides in the host (i.e., aphid) genome.[273,277] This observation is surprising since the endosymbiote possesses its own protein-synthetic machinery. All these findings were determined by using prokaryotic-specific antibiotics (e.g., rifampicin and chloramphenicol) to inhibit RNA and protein synthesis *in vitro*.[270,271] Moreover, the endosymbiotes are capable of the *de novo* biosynthesis of hundreds of "intrinsic" proteins when maintained *in vitro*. These proteins are not normally found in the endosymbiotes when in their host and their synthesis is presumably repressed by the host.[272] Based on these observations, Ishikawa concluded that protein synthesis in the endosymbiotes is under the influence of the host cell.

The use of both prokaryotic- and eukaryotic-specific antibiotics by Ishikawa in *in vivo* studies of pea aphid endosymbiotes shows a fascinating integration of host and endosymbiote protein biosynthesis. When the pea aphid is injected with cycloheximide (an antibiotic that affects only eukaryotic protein synthesis), only one 63 kDa protein is synthesized. Synthesis of this protein, designated as symbionin, is aborted when aphids are injected with chloramphenicol, a prokaryotic-specific antibiotic that prevents translation at the ribosome level. However, symbionin is still synthesized when aphids are injected with cycloheximide and rifampicin (rifampin), a prokaryotic-specific antibiotic which inhibits RNA polymerases.[274] Hence, Ishikawa concludes that the gene encoding for symbionin, its transcription, and synthesis of its mRNA occurs in the host cell. Translation and protein synthesis of symbionin occurs in the endosymbiote using symbiote tRNAs and ribosomes.[276] Ishikawa suggests that this sequence in protein synthesis by pea aphid endosymbiotes is comparable with protein synthesis in mitochondria and chloroplasts. Thus, he speculates that aphid endosymbiotes may be some ancestral or intermediate form of cell organelle within the mycetocyte.[280,282] Symbionin is also synthesized by endosymbiotes in the kondo aphid,[279,284] A. kondoi, a close relative of the pea aphid. Symbionin of the kondo aphid exhibits slightly different properties than pea aphid symbionin.

The biological function of symbionin is still unknown. It appears that it may have some function (perhaps nutritional) associated with aphid reproduction and embryonic development. Control of the biosynthesis of symbionin is age dependent. As viviparous aphids age, their ability to repress protein synthesis by their endosymbiotes diminishes. Concurrently, older aphids (45 to 50 d-old) no longer synthesize symbionin.[275,278] Pea aphids injected with rifampicin give birth to undersized offspring which are sterile and do not synthesize symbionin.[285] This is due to the removal or debilitation of the endosymbiotes by the antibiotic. It appears that symbionin synthesis is greatest during embryonic development.[287] During postembryonic development, the symbionin is metabolized and eventually depleted. The amount of symbionin produced in male aphids is significantly less than that produced in females. Furthermore, symbionin is found in overwintering eggs and, after hatching, it is the dominant protein found in the fundatrices.[286] It is important to note that the number of mycetocytes in pea aphids starts to decline from birth and reaches 0 in postreproductive individuals.[157] Hence, the depletion of symbionin may simply reflect the declining numbers of endosymbiotes and not result from the regulation of protein synthesis during aging.

The results of the above-mentioned molecular biology studies performed by Ishikawa are based on indirectly manipulating protein synthesis with antibiotics. Nonetheless, it is the view of Ishikawa that protein synthesis in the endosymbiotes is exceptionally limited. However, others have suggested that aphid endosymbiotes synthesize numerous other compounds (e.g., steroids, amino acids, polyketide pigments, etc., *vide supra*). The biosynthetic pathways involved in the synthesis of any of these other compounds involves many enzyme-catalyzed steps. Recently, Ishikawa estimated the genome size of the pea aphid endosymbiote to be $>1 \times 10^{10}$, several times larger than that of *E. coli*.[282] Perhaps future studies using recombinant DNA techniques will confirm the above results and answer some of the questions still remaining about biosynthetic capabilities of aphid endosymbiotes.

C. ENDOSYMBIOTES AND INSECT BIOTYPES

There is some rudimentary evidence that endosymbiotes may have a functional role in the formation of biotypes in insects, particularly with regard to aphid-host plant compatibility. Some indication that endosymbiotes may biosynthesize and supply enzymes for host-plant adaptation is provided by the mechanism(s) involved with biotype formation. Much of this hypothesis is conjectural.

Among the insects, aphids are renowned in their ability to form new, morphologically indistinguishable races that are referred to as biotypes.[164] Although biotype formation can be induced by pesticides, the most frequent reason for biotype formation is the adaptation to new host plants,[154,164] especially to new varieties of resistant crop plants.[153,154,164] Biotypes of aphids are often designated on the basis of their performance on a range of potential host plants.[164,305] Electrophoretic and/or morphometric analyses have not fully succeeded in showing distinct differences between host-plant-induced aphid biotypes. The ambiguity in the distinction between electromorphs of these aphids appears to be due to a high degree of genetic variation within and between biotypes and polymorphism at the enzyme loci studied.[33,353,507,566] Morphometric differences between these biotypes have been reported,[268,386] but the differences were only established with large sample sizes accompanied by fairly extensive statistical analyses. This would generally indicate that morphological distinctiveness between these biotypes is vague. Apart from performance differences on host plants, it has been reported that two biotypes of the greenbug also respond differently to scotophase-initiation of sexual morphs.[177]

With the exception of biotype formation due to pesticides,[150] the biochemical basis of aphid adaptation to new host plants has been obscure. There have been two schools of thought on the subject.[160] Some people believe that aphid adaptability to new host plants is associated with nutritional factors in the plant (e.g., the presence or absence of particular amino acids). Another belief is that host-plant resistance to aphids is related to toxic or deterring secondary plant compounds. However, the role of these particular plant chemical factors in resistance towards aphids and initiation of new biotypes is ambiguous.[160,409]

Recently, Dreyer and Campbell suggested that structural changes in matrix polysaccharides, either through breeding or induction by plant-growth regulators, can result in resistance of plants towards the prevailing aphid biotype within a given geographical area.[160] New biotypes of aphids that overcome this form of host-plant resistance would possess a new complement of salivary polysaccharases that enable them to depolymerize the altered matrix polysaccharides more effectively than their progenitors. It is not clear if the improved ability to depolymerize these polysaccharides is associated with facilitated penetration of the plant by the stylets of the aphid or if the breakdown products act as feeding cues. Oligosaccharides that result from the breakdown of plant matrix polysaccharides may signal host-plant suitability and aid in directing the stylets of the aphid towards the phloem.[85,90] Moreover, the oligosaccharides produced by the enzymatic breakdown of plant matrix polysaccharides are known elicitors of a variety of plant responses, including autotoxicity,[552] which is frequently encountered when aphids feed on a nonresistant host plant.

The plant pathology and biochemical mechanisms surrounding aphid-host plant interactions resemble those of plant-pathogen interactions (viz., secretion of polysaccharide-digesting enzymes which yield oligosaccharide elicitors that initiate autotoxicity or hypersensitive responses in the plant).[160] These observations have led to the speculation that the plant polysaccharide-degrading enzymes that are frequently found in sap-feeding insects are biosynthesized by their bacterial endosymbiotes.[86,160] If this is the case, then aphid biotype formation is partially a result of mutations or changes occurring in the endosymbiotes that are then passed on to the host. The possibility also exists that the endosymbiote genome may be altered by plasmids infecting the symbiotes. Because of parthenogenetic reproduction and the transovarial transmission of endosymbiotes in aphids, a large population of a better-adapted aphid biotype could appear very rapidly. This scenario is perhaps not all that far-fetched if one considers some recent findings. First, the sexual inheritance of host-plant compatibility traits which differentiate greenbug biotypes comes only from the mother aphid.[176] Moreover, the genes associated with these traits are extranuclear, indicating that they may be from the endosymbiotes of the mother aphid. Second, the rate of change in endosymbiote genes is much more rapid than it is in the aphid host.[284]

Biotype formation associated with different polysaccharases may not be confined to aphids. Two sibling species of *Sitophilus* grain weevils, *S. oryzae* and *S. zeamais*, can be completely differentiated based upon their polygalacturonase and pectinmethylesterase isozymes.[89] Similarly, nine host-plant races of weevils of *Rhynocyllus conicus* and *Bangasternus orientalis* can also be differentiated by the isozyme patterns of the same enzymes.[88] Because these weevils harbor endosymbiotes, it is possible that they are the source of the pectinases. Differences between endosymbiotes of these weevils could account for their differences in pectinase isozymes. Southern hybridizations of the Eco RI digests of the genes for 16S rRNAs of the endosymbiotes of *S. zeamais* and *S. oryzae* show distinctive differences.[91] It is reasonable to assume that their endosymbiotes are different. However, it is still necessary to prove that endosymbiotes are the source of pectinases in weevils as well as in aphids. Since there is presently no means of culturing these endosymbiotes, proof will come only after the successful cloning of these genes into a suitable, culturable bacterium that will express them *in vitro*.

IV. CONCLUSIONS AND SUMMARY

Although approximately 10% of all species of insects are known to harbor intracellular symbiotes, the biochemical and taxonomic characterization of these endosymbiotes has eluded definitive confirmation. One of the chief obstacles to their characterization is the absence of an artificial medium for their culture. Classification of insect endosymbiotes (especially prokaryotic) based upon their metabolism or ultrastructure has generally resulted in ambiguous designations. Use of histochemical or ultrastructural techniques has only resulted in similar ambiguities. Without a solid foundation in understanding either the biochemical or taxonomic nature of insect endosymbiotes, their biological contribution to insect survival can only be conjectured.

Much circumstantial evidence on the taxonomy and biochemistry of insect endosymbiotes has been garnered through research that focused on endosymbiotes within the confines of the host cell. Such studies attempted to manipulate the metabolism or genetic expression of either the host cell or symbiote with selective antibiotics or treatment (e.g., heat, lysozyme, etc.). However, because of the integral biochemical interaction between the symbiote and the host cell, the evidence should be interpreted cautiously.

Although there are still major gaps in our knowledge of insect endosymbiotes, there is a great deal known about the biochemistry, taxonomy, and physiological role of certain microorganisms that are extracellular symbiotes of insects. The relatively reliable information available about the nature of these extracellular symbiote-host relationships may stimulate insight on the role of endosymbiotes in insects.

Recent investigations of insect endosymbiotes have utilized genetic engineering techniques (e.g., recombinant DNA). These newer techniques have enabled researchers to study mechanisms of gene expression with respect to transcription and translation of host vs. symbiote DNA. Furthermore, they have been used to determine the nucleotide sequences of rRNA genes (vis., rDNA) of endosymbiotes. These sequences are highly conserved in the course of evolution and are reliable indicators of evolutionary relationships. Comparisons of nucleotide sequences of these genes are especially useful in elucidating the ancestral lineage of prokaryotic cells whose unique biochemical or physiological characteristics are difficult to distinguish or define. Comparing these sequences to free-living bacteria, using "nearest neighbor" cladistics, is currently the most promising approach towards characterizing the taxonomy of insect endosymbiotes. Moreover, taxonomic assignment of insect endosymbiotes will furnish some indication of their biochemical characteristics as well.

Insect endosymbiotes may be synthesizing compounds that play a role in the physiological or nutritional status of the host insect. There is a strong indication that these symbiotes are providing the host with enzymes that break down plant matrix heteropolysaccharides (a role that has been readily invoked for extracellular gut symbiotes). These heteropolysaccharides are generally refractory to hydrolytic enzymes normally synthesized by eukaryotic (i.e., host) cells. The production of depolymerizing enzymes or the biosynthesis of essential nutrients by these endosymbiotes would have a significant impact on insect feeding and/or nutrition and the evolution of insect-plant interactions in general. Such processes by endosymbiotes may be integral to biotype formation in insects. However, there is still a great deal to learn about the biochemistry and classification of insect symbiotes. It remains to be seen if there is some prospect of exploiting the genetic information of insect endosymbiotes for application in insect pest management.

ACKNOWLEDGMENTS

I wish to thank the following for their helpful suggestions and review of the manuscript: R. Binder, J. Steffen, and D. Light at USDA-ARS, WRRC, Albany, CA; P. Baumann and B. Unterman at the Department of Bacteriology, University of California, Davis; and R. Chapman and E. Bernays at the Department of Entomology, University of California, Berkeley. A special recognition goes to David L. Dreyer, WRRC (retired), who inspired much of the current research on the chemical basis of interactions between plants and aphids (and their endosymbiotes).

REFERENCES

1. **Abrahamson, L. P. and Norris, D. M.,** Symbiotic interrelationship between microbes and ambrosia beetles. IV. Ambrosial fungi associated with *Xyloterinus politus, J. Invertebr. Pathol.,* 14, 381, 1969.
2. **Akey, D. H. and Beck, S. D.,** Continuous rearing of the pea aphid, *Acyrthosiphon pisum,* on a holidic diet, *Ann. Entomol. Soc. Am.,* 64, 353, 1971.
3. **Akey, D. H. and Beck, S. D.,** Nutrition of the pea aphid, *Acyrthosiphon pisum:* requirements for trace metals, sulphur and cholesterol, *J. Insect Physiol.,* 18, 1901, 1972.
4. **Allen, T. C.,** Bacteria producing rot of apple in association with the apple maggot, *Rhagoletis pomonella, Phytopathology,* 21, 338, 1931.
5. **Allen, T. C. and Riker, A. J.,** A rot of apple fruit caused by *Phytomonas melophthora* following invasion by the apple maggot, *Phytopathology,* 22, 557, 1932.

6. **Allen, T. C., Pinckard, J. A., and Riker, A. J.,** Frequent association of *Phytomonas melophthora* with various stages in the life cycle of the apple maggot, *Rhagoletis pomonella, Phytopathology,* 24, 228, 1934.
7. **Amiressami, M.,** Investigation of the light microscopical and ultrastructure of the Demeton-S-methyl resistance aphids under consideration of the mycetome symbionts of the *Phorodon humuli* Schrank, in *Endocytobiology,* Vol. 1, Schwemmier, W. and Schenk, H. E. A., Eds., Walter de Gruyter, Berlin, 1980, 425.
8. **Amiressami, M.,** Ultrastructure relationships of intracellular bacteria-like micro-oganisms during embryogenesis of *Ornithodorus moubata* Murray (Ixodoidea: Argassidae), in *Endocytobiology,* Vol. 2, Schenk, H. E. A. and Schwemmier, W., Eds., Walter de Gruyter, Berlin, 1983, 775.
9. **Amiressami, M. and Petzoid, H.,** Licht- und elektronemikroskopische Untersuchungen über das Verhalten der Mycetomsymbionten bei insektizidresistenten und Normal-Sensibien Pfirsichblattläusen *(Myzus persicae* Sulz.), *Z. Angew. Zool.,* 3, 273, 1976.
10. **Amiressami, M. and Petzoid, H.,** Symbioseforschung und Insektizidresistenz: ein Licht- und Elektronenmikroskopischer Beitrag zur Klŕung der Insektizidresistenz von Aphiden unter Berücksichtigung der Mycetomsymbionten bei *Myzus persicae* Sulz., *Z. Angew. Entomol.,* 82, 252, 1977.
11. **Andrews, R. E. and Spence, K. D.,** Action of Douglas fir tussock moth larvae and their microflora on dietary terpenes, *Appl. Environ. Microbiol.,* 40, 959, 1980.
12. **Aschner, M.,** Experimentalle Untersuchungen über die Symbiose der Kleiderlaus, *Naturwissenschaften,* 20, 501, 1932.
13. **Aschner, M.,** Studies on the symbiosis of the body-louse. I. Elimination of the symbionts by centrifugalisation of the eggs, *Parasitology,* 26, 309, 1934.
14. **Aschner, M. and Ries, E.,** Das Verhalten der Kleiderlaus bei Ausschaltung ihrer Symbionten. Eine Experimentelle Symbiosstudie, *Z. Morphol. Öekol. Tiere,* 26, 529, 1933.
15. **Auclair, J. L.,** Aphid feeding and nutrition, *Annu. Rev. Entomol.,* 8, 439, 1963.
16. **Auclair, J. L.,** Feeding and nutrition of the pea aphid, *Acyrthosiphon pisum* (Homoptera: Aphididae), on chemically defined diets of various pH and nutrient levels, *Ann. Entomol. Soc. Am.,* 58, 855, 1965.
17. **Auclair, J. L.,** Effects of pH and sucrose on rearing the cotton aphid, *Aphis gossypii* on germ-free and holidic diet, *J. Insect Physiol.,* 13, 431, 1967.
18. **Auclair, J. L.,** Nutrition of plant sucking insects on chemically defined diets, *Entomol. Exp. Appl.,* 13, 307, 1969.
19. **Auclair, J. L. and Cartier, J. J.,** Pea aphid: rearing on a chemically defined diet, *Science,* 142, 1068, 1963.
20. **Aucliar, J. L. and Srivastava, P. N.,** Some mineral requirements of the pea aphid, *Acyrthosiphon pisum, Can. Entomol.,* 104, 927, 1969.
21. **Baervald, R. J. and Boush, G. M.,** Demonstration of the bacterial symbiote *Pseudomonas melophthora* in the apple maggot, *Rhagoletis pomonella,* by fluorescent-antibody technique, *J. Invertebr. Pathol.,* 11, 251, 1968.
22. **Baines, S.,** Bacterial symbiosis in the assassin bug, *J. Gen. Microbiol.,* 8, iv, 1953.
23. **Baines, S.,** The role of the symbiotic bacteria in the nutrition of *Rhodnius prolixus* (Hemiptera), *J. Exp. Biol.,* 33, 533, 1956.
24. **Bak, A. L.,** DNA base composition in *Mycoplasma,* bacteria and yeast, *Curr. Top. Microbiol. Immunol.,* 61, 89, 1973.
25. **Batra, L. R., Ed.,** *Insect-Fungus Symbiosis. Nutrition, Mutualism, and Commensalism,* John Wiley & Sons, New York, 1979, 276.
26. **Batra, L. R. and Batra, W. S.,** Termite-fungus mutualism, in *Insect-Fungus Symbiosis. Nutrition, Mutualism, and Commensalism,* Batra, L. R., Ed., John Wiley & Sons, New York, 1979, 117.
27. **Bayon, C.,** Transit des aliments et fermentations continues dans le tube digestif d'une larvae xylophage d'insecte: *Oryctes nasicornis* (Coleoptera: Scarabaeidae), *C. R. Acad. Sci. Paris,* 290, 1145, 1980.
28. **Bayon, C.,** Volatile fatty acids and methane production in relation to anaerobic carbohydrate fermentation in *Oryctes nasicornis* larvae (Coleoptera: Scarabaeidae), *J. Insect Physiol.,* 26, 819, 1980.
29. **Bayon, C. and Etievant, P.,** Methanic fermentation in the digestive tract of a xylophagous insect: *Oryctes nasicornis* L. larva (Coleoptera: Scarabaeidae), *Experientia,* 36, 154, 1980.
30. **Bayon, C. and Mathelin, J.,** Carbohydrate fermentation and by-product absorption studied with labelled cellulose in *Oryctes nasicornis* larvae (Coleoptera: Scarabaeidae), *J. Insect Physiol.,* 26, 833, 1980.
31. **Behrenz, W. and Technau, G.,** Versuche zur Bekampfung von *Anobium punctatum* mit Symbionticiden, *Z. Angew. Entomol.,* 44, 22, 1959.
32. **Bentley, B. L.,** Nitrogen fixation in termites: fate of newly fixed nitrogen, *J. Insect Physiol.,* 30, 653, 1984.
33. **Beregovoy, V. H. and Starks, K. J.,** Enzyme patterns in biotypes of the greenbug, *Schizaphis graminum* (Rondani) (Homoptera: Aphididae), *J. Kansas Entomol. Soc.,* 59, 517, 1986.
34. **Bhatnagar, R. D. S. and Musgrave, A. J.,**

Cytochemistry, morphogenesis, and tentative identification of mycetomal microorganisms of *Sitophilus granarius* L. (Coleoptera), *Can. J. Microbiol.,* 16, 1357, 1970.

35. **Bignell, D. E.,** Some observations on the distribution of gut flora in the American cockroach, *Periplaneta americana, J. Invertebr. Pathol.,* 29, 338, 1977.
36. **Bignell, D. E.,** An ultrastructural study and stereological analysis of the colon wall in the cockroach *Periplaneta americana, Tissue Cell,* 12, 153, 1980.
37. **Bignell, D. E., Oskarsson, H., and Anderson, J. M.,** Association of actinomycete-like bacteria with soil-feeding termites (Termitidae, Termitinae), *Appl. Environ. Microbiol.,* 37, 339, 1979.
38. **Bignell, D. E., Oskarsson, H., and Anderson, J. M.,** Colonization of the epithelial face of the peritrophic membrane and the ectoperitrophic space by actinomycetes in a soil-feeding termite, *J. Invertebr. Pathol.,* 36, 426, 1980.
39. **Bignell, D. E., Oskarsson, H., and Anderson, J. M.,** Distribution and abundance of bacteria in the gut of a soil-feeding termite *Procubitermes aburiensis* (Termitidae, Termitinae), *J. Gen. Microbiol.,* 117, 393, 1980.
40. **Bignell, D. E., Oskarsson, H., and Anderson, J. M.,** Specialization of the hindgut wall for the attachment of symbiotic microorganisms in a termite *Procubitermes aburiensis* (Isoptera, Termitidae, Termitinae), *Zoomorphology,* 96, 103, 1980.
41. **Blewett, N. and Frainkel, G.,** Intracellular symbiosis and vitamin requirements of two insects, *Lasioderma serricorne* and *Sitodrepa panicea, Proc. R. Soc. London Ser. B,* 132, 212, 1944.
42. **Blomquist, G. J., Dwyer, L. A., Chu, A. J., Ryan, R. O., and de Renobales, M.,** Biosynthesis of linoleic acid in a termite, cockroach and cricket, *Insect Biochem.,* 12, 349, 1982.
43. **Bloodgood, R. A. and Fitzharris, T. P.,** Specific association of prokaryotes with symbiotic flagellate Protozoa from the hindgut of the termite *Reticulitermes* and the wood-eating roach *Cryptocerus, Cytobios,* 17, 103, 1976.
44. **Blum, M. S.,** Biosynthesis of arthropod exocrine compounds, *Annu. Rev. Entomol.,* 32, 381, 1987.
45. **Borden, J. H., Ryker, L. C., Chong, L. J., Pierce, H. D., Jr., Johnston, B. D., and Oehlschlager, A. C.,** Response of the mountain pine beetle, *Dendroctonus ponderosae* Hopkins (Coleoptera: Scolytidae), to five semiochemicals in British Columbia lodgepole pine forests, *Can. J. For. Res.,* 17, 118, 1987.
46. **Botha, T. C. and Hewitt, P. H.,** A study of the gut morphology and some physiological observations on the influence of diet of green *Themeda triandra* on the harvester termite, Hodotermes mossambicus (Hagen), *Phytophylactica,* 11, 57, 1979.
47. **Boush, G. M. and Coppel, H. C.,** Symbiology: mutualism between arthropods and microorganisms, in *Insect Diseases,* Vol. 1, Cantwell, G. E., Ed., Marcel Dekker, New York, 1974, 301.
48. **Boush, G. M. and Matsumura, F.,** Insecticidal degradation by *Pseudomonas melophthora,* the bacterial symbiote of the apple maggot, *J. Econ. Entomol.,* 60, 918, 1967.
49. **Boush, G. M., Saleh, S. M., and Baranoski, R. M.,** Bacteria associated with the Caribbean fruit fly, *Environ. Entomol.,* 1, 30, 1972.
50. **Boyd, N. M. and Martin, M. M.,** Faecal proteinases of the fungus-growing ant, *Atta texana:* their fungal origin and ecological significance, *J. Insect Physiol.,* 21, 1815, 1975.
51. **Bracke, J. W. and Markovetz, A. J.,** Immunolatex localization by scanning electron microsopy of intestinal bacteria from cockroaches, *Appl. Environ. Microbiol.,* 38, 945, 1979.
52. **Bracke, J. W., Cruden, D. L., and Markovetz, A. J.,** Intestinal microbial flora of the American cockroach, *Periplaneta americana* L., *Appl. Environ. Microbiol.,* 38, 945, 1979.
53. **Brand, J. M. and Barras, S. J.,** The major volatile constituent of a basidiomycete associated with the Southern pine beetle, *Lloydia,* 40, 398, 1977.
54. **Brand, J. M., Bracke, J. W., Britton, L. N., Markovetz, A. J., and Barras, S. J.,** Bark beetle pheromones: production of verbenone by a mycangial fungus of *Dendroctonus frontalis, J. Chem. Ecol.,* 2, 195, 1976.
55. **Brand, J. M., Bracke, J. W., Markovetz, A. J., and Browne, L. E.,** Production of verbenol pheromone by a bacterium isolated from bark beetles, *Nature (London),* 254, 136, 1975.
56. **Brand, J. M., Schultz, J., Barras, S. J., Edson, L. J., Payne, T. L., and Hedden, R. L.,** Bark beetle pheromones: enhancement of *Dendroctonus frontalis* (Coleoptera: Scolytidae) aggregation pheromone by yeast metabolites in laboratory bioassays, *J. Chem. Ecol.,* 3, 657, 1977.
57. **Brand, J. M., Young, J. C., and Silverstein, R. M.,** Insect pheromones: a critical review of recent advances in their chemistry, biology, and application, *Fortschr. Chem. Org. Naturst.,* 37, 112, 1979.
58. **Brecher, G. and Wigglesworth, V. B.,** The

transmission of *Actinomyces rhodnii* Erikson in *Rhodnius prolixus* and its influence on the growth of the host, *Parasitology,* 35, 220, 1944.

59. **Breznak, J. A.,** Intestinal microbiota of termites and other xylophagous insects, *Annu. Rev. Microbiol.,* 36, 323, 1982.
60. **Breznak, J. A.,** Biochemical aspects of symbiosis between termites and their intestinal microbiota, in *Invertebrate-Microbial Interactions,* Anderson, J. M., Rayner, A. D. M., and Walton, D., Eds., Cambridge University Press, Cambridge, 1983.
61. **Breznak, J. A. and Pankratz, H. S.,** *In situ* morphology of the gut microbiota of wood-eating termites [*Reticulitermes flavipes* (Kollar) and *Coptotermes formosanus* Shiraki], *Appl. Environ. Microbiol.,* 33, 406, 1977.
62. **Breznak, J. A., Brill, W. J., Mertins, J. W., and Coppel, H. C.,** Nitrogen fixation in termites, *Nature (London),* 244, 577, 1973.
63. **Bridges, J. R.,** Nitrogen-fixing bacteria associated with bark beetles, *Microbiol. Ecol.,* 7, 131, 1981.
64. **Briscoe, M. S.,** A survey of the protozoan fauna of the cockroach, *Blaberus craniifer, J. Invertebr. Pathol.,* 17, 291, 1971.
65. **Bromel, M., Duh, F. M., Erdmann, G. R., Hammack, L., and Gassner, G.,** Bacteria associated with the screwworm fly [*Cochliomyia homnivorax* (Coquerel)] and their metabolites, in *Endocytobiology II, Intracellular Space as Oligogenetic Ecosystem,* Schenk, E. A. and Schwemmier, W., Eds., Walter de Gruyter, Berlin, 1983, 791.
66. **Brooks, M. A.,** Nature and significance of intracellular bacteroids in cockroaches, *Proc. Int. Congr. Entomol.,* 10, 311, 1956.
67. **Brooks, M. A.,** Symbiosis and aposymbiosis in arthropods, *Symp. Soc. Gen. Microbiol.,* 13, 200, 1963.
68. **Brooks, M. A.,** Comments on the classification of intracellular symbiotes of cockroaches and a description of the species, *J. Invertebr. Pathol.,* 16, 249, 1970.
69. **Brooks, M. A. and Richards, A. G.,** Intracellular symbiosis in cockroaches. I. Production of aposymbiotic cockroaches, *Biol. Bull.,* 109, 22, 1955.
70. **Brooks, M. A. and Richards, A. G.,** Intracellular symbiosis in cockroaches. II. Mitotic division of mycetocytes, *Science,* 122, 242, 1955.
71. **Brooks, M. A. and Richards, K.,** On the *in vitro* culture of intracellular symbiotes of cockroaches, *J. Invertebr. Pathol.,* 8, 150, 1966.
72. **Brown, K. S., Jr.,** The chemistry of aphids and scale insects, *Chem. Soc. Rev. London,* 4, 263, 1975.
73. **Brown, K. S., Jr. and Weiss, U.,** Chemical constituents of the bright orange aphid, *Aphis nerii* Fonscolombe. II. Structures of some minor components, *An. Acad. Bras. Ciěnc.,* Suppl. 42, 205, 1970.
74. **Brown, K. S., Jr. and Weiss, U.,** Chemical constituents of the bright orange aphid, *Aphis nerii* Fonscolombe. III. Two naphthaquinone methides of unusual structure and chemical behavior, *Tetrahedron Lett.,* 38, 3501, 1971.
75. **Brown, K. S., Jr., Cameron, D. W., and Weiss, U.,** Chemical constituents of the bright orange aphid, *Aphis nerii,* Fonscolombe. I. Neriaphin and 6-hydroxymusizin 8-*o*-b-D-glucoside, *Tetrahedron Lett.,* 6, 471, 1969.
76. **Buchner, P.,** *Tier und Pflanze in intracellularer Symbiose,* Gerbrüder Borntraeger, Berlin, 1921, 462.
77. **Buchner, P.,** Endosymbiosestudien on Schildlausen. IV. *Hippeococcus,* eine Myrmekophile Pseudococcine, *Z. Morphol. Öekol. Tiere,* 45, 379, 1957.
78. **Buchner, P.,** Eine neue Form der Endosymbiose bei Aphiden, *Zool. Anz.,* 160, 222, 1958.
79. **Buchner, P.,** *Endosymbiose der Tiere mit pflanzlichen Mikroorganismen,* Birkhauser, Basel, 1953, 771; Engl. trans., John Wiley & Sons, New York, 1965, 901.
80. **Burgerdorfer, W., Brinton, L. P., and Hughes, L. E.,** Isolation and characterization of symbiotes from the Rocky Mountain wood tick, *Dermacentor andersoni, J. Invertebr. Pathol.,* 22, 424, 1973.
81. **Burges, H. D., Grove, J. R., and Pople, M.,** The internal microbial flora of the elm bark beetle, *Scolytus scolytus,* at all stages of development, *J. Invertebr. Pathol.,* 34, 21, 1979.
82. **Bush, G. L. and Chapman, G. B.,** Electron microscopy of symbiotic bacteria in developing oocytes of the American cockroach, *Periplaneta americana, J. Bacteriol.,* 81, 267, 1961.
83. **Butler, J. H. A. and Buckerfield, J. C.,** Digestion of lignin by termites, *Soil Biol. Biochem.,* 11, 507, 1979.
84. **Byers, J. A. and Wood, D. L.,** Antibiotic-induced inhibition of pheromone synthesis in a bark beetle, *Science,* 213, 763, 1981.
85. **Campbell, B. C.,** Host-plant oligosaccharins in the honeydew of *Schizaphis graminum* (Rondani) (Insecta, Aphididae), *Experientia,* 42, 451, 1986.
86. **Campbell, B. C. and Dreyer, D. L.,** Host-plant resistance of sorghum: differential hydrolysis of pectic substances by polysaccharases of greenbug biotypes *(Schizaphis graminum,* Homoptera, Aphididae), *Arch. Insect Biochem. Physiol.,* 2, 203, 1985.
87. **Campbell, B. C. and Nes, W. D.,** A reap-

praisal of sterol biosynthesis and metabolism in aphids, *J. Insect Physiol.*, 29, 149, 1983.

88. **Campbell, B. C., Turner, C., and Maddox, D.**, unpublished data, 1986.
89. **Campbell, B. C. and Van Fleet, J.**, unpublished data, 1986.
90. **Campbell, B. C., Jones, K. C., and Dreyer, D. L.**, Discriminative behavioral responses by aphids to various plant matrix polysaccharides, *Entomol. Exp. Appl.*, 41, 17, 1986.
91. **Campbell, B. C., Unterman, B., and Baumann, P.**, unpublished data, 1987.
92. **Carpenthier, R., Carpenthier, B., and Zethner, O.**, The bacterial flora of the midgut of two Danish populations of healthy fifth instar larvae of the turnip moth, *Scotia segetum, J. Invertebr. Pathol.*, 32, 59, 1978.
93. **Carter, W.**, The symbionts of *Pseudococcus brevipes* (Ckl.), *Ann. Entomol. Soc. Am.*, 28, 60, 1935.
94. **Carter, W.**, The symbionts of *Pseudococcus brevipes* in relation to a phyto-toxic secretion of the insect, *Phytophathology,* 26, 176, 1936.
95. **Chang, K. P.**, Effects of elevated temperature on the mycetome and symbiotes of the bed bug *Cimex lectularis, J. Invertebr. Pathol.*, 23, 333, 1974.
96. **Chang, K. P.**, Haematophagous insect and haemoflagellate as hosts for prokaryotic endosymbionts, *Symp. Soc. Exp. Biol.*, 29, 407, 1975.
97. **Chang, K. P. and Musgrave, A. J.**, Histochemistry and ultrastructure of the mycetome and its 'symbiotes' in the pear psylla, *Psylla pyricola* Foerster, *Tissue Cell*, 1, 597, 1969.
98. **Chang, K. P. and Musgrave, A. J.**, Ultrastructure of rickettsia-like microorganisms in the midgut of a plant bug, *Stenotus binotatus* Jak., *Can. J. Microbiol.*, 16, 621, 1970.
99. **Chang, K. P. and Musgrave, A. J.**, Multiple symbiosis in a leafhopper, *Helochara communis* Fitch: envelcpes, nucleoids and inclusions of the symbiotes, *J. Cell Sci.*, 11, 275, 1972.
100. **Chang, K. P. and Musgrave, A. J.**, Conversion of spheroplast symbiotes in a leafhopper, *Helochara communis* Fitch, *Can. J. Microbiol.*, 21, 196, 1975.
101. **Chang, K. P. and Musgrave, A. J.**, Endosymbiosis in a leafhopper, *Helochara communis* Fitch: symbiote translocation and auxiliary cells in the mycetome, *Can. J. Microbiol.*, 21, 186, 1975.
102. **Chang, K. P. and Musgrave, A. J.**, Morphology, histochemistry and ultrastructure of mycetome and its rickettsial symbiotes in
103. **Chapman, R. F.**, *The Insects: Structure and Function,* Elsevier, New York, 1975, 57.
104. **Chararas, C.**, Physiologie des insectes. L'action synergique des constituants glucidiques et des constituants terpéniques dans le processus d'attraction secondaire et le mécanisme de l'élaboration des phéromones chez le Scolytidae parasites des confères, *C. R. Acad. Sci.*, 284, 1545, 1977.
105. **Chararas, C.**, *Ecophysiologie des Insectes Parasites des Foréts,* Published by author, Paris, 1980, 297.
106. **Chararas, C.**, Etude du comportement nutritionnel et de la digestion chez certains Cerambycidae xylophages, *Mater. Org.*, 16, 207, 1981.
107. **Chararas, C. and Libols, G.**, Physiologie des insectes. Étude des enzymes hydrolisant les osides la larve d'*Ergates faber* L. (Coléoptère Cerambycidae), *C. R. Acad. Sci. Paris,* 283, 1523, 1976.
108. **Chararas, C., Chipoulet, J. M., and Courtois, J. E.**, Etude du préférendum alimentaire et des osidases de *Pyrochroa coccinea* (Coléoptère Pyrochroidae), *C. R. Séances Soc. Biol.*, 173, 42, 1979.
109. **Chararas, C., Chipoulet, J. M., and Courtois, J. E.**, Purification partielle et caractérisation d'une β-glucosidase des larves de *Pyrochora coccinea* (Coléoptère Pyrochroidae), *C. R. Séances Soc. Biol.*, 177, 22, 1983.
110. **Chararas, C., Courtois, J. E., and Laurent-Hubé, H.**, Influence de l'hamycine sur les larves xylophages de *Capnodis* et leurs osidases, *Ann. Pharm. Fr.*, 25, 257, 1967.
111. **Chararas, C., Courtois, J. E., Debris, M.-M., and Laurent-Hubé, H.**, Activités comparées des osidases chez divers stades de deux insectes xylophages, parasites de confirères, *Bull. Soc. Chim. Biol.*, 45, 383, 1963.
112. **Chararas, C., Courtois, J. E., Thuillier, A., Le Fay, A., and Laurent-Hubé, H.**, Nutrition de *Phoracantha semipunctata* F. (Coléoptère Cerambicidae); étude des osidases du tube digestif et de las flore intestinale, *C. R. Séances Soc. Biol.*, 166, 304, 1972.
113. **Chararas, C., Eberhard, R., Courtois, J. E., and Petek, F.**, Purification of three cellulases from the xylophagous larvae of *Ergates faber* (Coleoptera: Cerambycidae), *Insect Biochem.*, 13, 213, 1983.
114. **Chararas, C., Pignal, M. C., Vodjani, G., and Bourgeaycausse, M.**, Glycosidases and B-group vitamins produced by 6 yeast strains from the digestive tract of *Phorancantha semipunctata* larvae and their role in the insect development, *Mycopathologia*, 83, 9, 1983.
115. **Chararas, C., Riviere, J., Ducause, C., Rutledge, D., Delpul, G., and Cazelles, M.-T.**, Physiologie des invertebres — bio-

conversion d'une composé terpenique sous action d'une bacterie du tube digestif de *Phloesinus armatus* (Coléoptère, Scolytidae), *C. R. Acad. Sci. Paris,* 291, 299, 1980.

116. **Chipoulet, J. M. and Chararas, C.,** Purification and partial characterization of a laminarinase from the larvae of *Rhagium inquisitor, Comp. Biochem. Physiol.,* 77B, 699, 1984.
117. **Chipoulet, J. M. and Chararas, C.,** Purification and partial characterization of a cellobiase from the larvae of *Rhagium inquistor, Comp. Biochem. Physiol. B,* 82B, 327, 1985.
118. **Chipoulet, J.-M. and Chararas, C.,** Survey and electrophoretical separation of the glycosidases of *Rhagium inquisitor* (Coleoptera: Cerambycidae) larvae, *Comp. Biochem. Physiol.,* 80B, 241, 1985.
119. **Clayton, R. B.,** The utilization of sterols by insects, *J. Lipid Res.,* 5, 3, 1964.
120. **Cleveland, L. R.,** The feeding habit of termite castes and its relation to their intestinal flagellates, *Biol. Bull.,* 48, 295, 1925.
121. **Cleveland, L. R.,** The effect of oxygenation and starvation on the symbiosis between the termite, *Termopsis,* and its intestinal flagellates *Biol. Bull.,* 48, 309, 1925.
122. **Cleveland, L. R.,** Symbiosis among animals with special reference to termites and their intestinal flagellates, *Q. Rev. Biol.,* 1, 51, 1926.
123. **Cleveland, L. R., Hall, S. R., Sanders, E. P., and Collier, J.,** The woodfeeding roach, *Cryptocercus,* its protozoa, and the symbiosis between protozoa and roach, *Mem. Am. Acad. Sci.,* 17, 185, 1934.
124. **Cochran, D. G.,** Nitrogen excretion in cockroaches, *Annu. Rev. Entomol.,* 30, 29, 1985.
125. **Cornwell, P. B.,** *The Cockroach,* Vol. 1, Hutchison and Co., London, 1968.
126. **Courtice, A. C. and Drew, R. A. I.,** Bacterial regulation of abundance in tropical fruit flies (Diptera: Tephritidae), *Aust. Zool.,* 21, 251, 1984.
127. **Courtois, J. E. and Chararas, C.,** Les enzymes hydrolysant les glucides (hydrates de carbone) chez les insectes xylophage parasites des conifères et de quelques autres arbres forestier, *Mater. Org.,* 1, 127/150, 1965.
128. **Courtois, J. E., Chararas, C., and Debris, M.-M.,** Étude de l'attaque enzymatique des glucides par un Coléoptère xylophage: *Ips typographus* L., *C. R. Acad. Sci. Paris,* 252, 2608, 1961.
129. **Courtois, J. E., Chararas, C., Debris, M. M., and Laurent-Hubé, H.,** Répartition comparée des osidases chez les insectes xylophages parasites des arbres forestiers, *Bull. Soc. Chim. Biol.,* 47, 2219, 1965.
130. **Cowdry, E. V.,** The distribution of *Rickettsia* in tissues of insects and arachnids, *J. Exp. Med.,* 37, 431, 1923.
131. **Crawford, R. E., McDermott, L. A., and Musgrave, A. J.,** Microbial isolations from the granary weevil *Sitophilus granarius* L. (Coleoptera: Curculionidae), *Can. Entomol.,* 92, 577, 1960.
132. **Cruden, D. L. and Markovetz, A. J.,** Carboxymethylcellulose decomposition by intestinal bacteria of cockroaches, *Appl. Environ. Microbiol.,* 38, 369, 1979.
133. **Cruden, D. L. and Markovetz, A. J.,** A thick-walled organism isolated from the cockroach gut by using a spent medium technique, *Appl. Environ. Microbiol.,* 39, 261, 1980.
134. **Cruden, D. L. and Markovetz, A. J.,** Relative numbers of selected bacterial forms in different regions of the cockroach hindgut, *Arch. Microbiol.,* 129, 129, 1981.
135. **Cruden, D. L. and Markovetz, A. J.,** Microbial aspects of the cockroach hindgut, *Arch. Microbiol.,* 138, 131, 1984.
136. **Cruden, D. L., Gorrell, T. E., and Markovetz, A. J.,** Novel microbial and chemical components of a specific blackband region in the cockroach hindgut, *J. Bacteriol.,* 140, 687, 1979.
137. **Czolij, R., Slaytor, M., and O'Brien, R. W.,** Bacterial flora of the mixed segment and the hindgut of the higher termite *Nasutitermes exitiosus* Hill (Termitidae, Nasutitermitinae), *Appl. Environ. Microbiol.,* 49, 1226, 1985.
138. **Dadd, R. H.,** Dietary amino acids and wing determination in the aphid *Myzus persicae, Ann. Entomol. Soc. Am.,* 61, 1201, 1968.
139. **Dadd, R. H. and Krieger, D. L.,** Continuous rearing of aphids of the *Aphis fabae* complex on sterile synthetic diet, *J. Econ. Entomol.,* 60, 1512, 1967.
140. **Dadd, R. H. and Krieger, D. L.,** Dietary amino acid requirements of the aphid, *Myzus persicae, J. Insect Physiol.,* 14, 741, 1968.
141. **Dadd, R. H. and Mittler, T. E.,** Studies on the artificial feeding of the aphid *Myzus persicae* (Sulzer). III. Some major nutritional requirements, *J. Insect Physiol.,* 11, 717, 1965.
142. **Dadd, R. H. and Mittler, T. E.,** Permanent culture of an aphid on a totally synthetic diet, *Experientia,* 22, 832, 1966.
143. **Dadd, R. H., Krieger, D. L., and Mittler, T. E.,** Studies on the artificial feeding of the aphid *Myzus persicae* (Sulzer). IV. Requirements for water-soluble vitamins and ascorbic acid, *J. Insect Physiol.,* 13, 249, 1967.
144. **Dang-Gabrani, K.,** On the functions of intracellular symbiotes of *Sitophilus oryzae* Linn., *Experientia,* 27, 107, 1971.
145. **Dasch, G. L.,** Morphological and Molecular

Studies on Intracellular Bacterial Symbiotes of Insects, Ph.D. thesis, Yale University, New Haven, CT, 1971.

146. **De Picels Polver, P., Sacchi, L., Grigolo, A., and Laudni, U.,** Fine structure of the body and its bacteroids in *Blattella germanica* (Blattoidea), *Acta Zool.,* 67, 63, 1986.

147. **Dean, R. W. and Chapman, P. J.,** Bionomics of the apple maggot in eastern New York, *Search Agric. Entomol. Geneva, N.Y.,* 3, 62, 1973.

148. **von Dehn, M.,** Untersuchungen über die Stoffwechselphysiologie der Aphiden. Die Aminosäuren und Zucker Siebröhrensaft einiger Krautgewächsarten und im Honigtau ihrer Schmarotzer, Z. Vgl. Physiol., 45, 88, 1961.

149. **Deschamps, P.,** Contribution à l'étude de la xylophagie. La nutrition des larves de Cérambycides, *Ann. Sci. Nat. Zool. Biol. Anim.,* 15, 449, 1953.

150. **Devonshire, A. L.,** The properties of a carboxylesterase from the peach-potato aphid, *Myzus persicae* (Sulz.), and its role in conferring insecticide resistance, *Biochem. J.,* 167, 675, 1977.

151. **Devonshire, A. L., Searle, L. M., and Moores, G. D.,** Quantitative and qualitative variation in the mRNA for carboxylesterases in insecticide-susceptible and resistant *Myzus persicae* (Sulz.), *Insect Biochem.,* 16, 659, 1986.

152. **Dickman, A.,** Studies on the wax moth, *Galleria mellonella,* with particular reference to the digestion of wax by the larvae, *J. Cell. Comp. Physiol.,* 3, 223, 1933.

153. **Diehl, S. R. and Bush, G. L.,** An evolutionary and applied perspective of insect biotypes, *Annu. Rev. Entomol.,* 29, 471, 1984.

154. **Dixon, A. F. G.,** Structure of aphid populations, *Annu. Rev. Entomol.,* 30, 155, 1985.

155. **Donnellan, J. F. and Kirby, B. A.,** Uric acid metabolism by symbiotic bacteria from the fat body of *Periplaneta americana, Comp. Biochem. Biophys.,* 22, 235, 1967.

156. **Douglas, A. E.,** On the source of sterols in the green peach aphid, *Myzus persicae,* reared on holidic diets, *J. Insect Physiol.,* in press.

157. **Douglas, A. E. and Dixon, A. F. G.,** The mycetocyte symbiosis of aphids: variation with age and morph in virginoparae of *Megoura viciae* and *Acyrthosiphon pisum, J. Insect Physiol.,* 33, 109, 1987.

158. **Downer, R. G. H.,** Functional role of lipids in insects, in *Biochemistry of Insects,* Rockstein, M., Ed., Academic Press, New York, 1978, 57.

159. **Drew, R. A. I., Courtice, A. C., and Teakle, D. S.,** Bacteria as a natural source of food for adult fruit flies (Diptera: Tephritidae), *Oecologia,* 60, 279, 1983.

160. **Dreyer, D. L. and Campbell, B. C.,** Chemical basis of host-plant resistance to aphids, *Plant Cell Environ.,* 10, 353, 1987.

161. **Dreyer, D. L., Reese, J. C., and Jones, K. C.,** Aphid feeding deterrents in sorghum: bioassay, isolation and characterization, *J. Chem. Ecol.,* 7, 273, 1981.

162. **Dubos, R. and Kessler, A.,** *Symbiotic Associations,* 13th Symp. Soc. General Microbiology, Cambridge, University Press, 1963.

163. **Duffey, J. E. and Powell, R. D.,** Microbial induced ethylene synthesis as a possible factor of square abscission and stunting in cotton infested by cotton fleahopper, *Ann. Entomol. Soc. Am.,* 72, 599, 1979.

164. **Eastop, V. F.,** Biotypes of aphids, in *Perspectives in Aphid Biology,* Lowe, A. D., Ed., *Bull. Entomol. Soc. N.Z.,* 2, 40, 1975.

165. **Eberle, M. W. and McLean, D. L.,** Initiation and orientation of the symbiote migration in the human body louse *Pediculus humanus* L., *J. Insect Physiol.,* 28, 417, 1982.

166. **Eberle, M. W. and McLean, D. L.,** Observation of symbiote migration in human body lice with scanning and transmission electron microscopy, *Can. J. Microbiol.,* 29, 755, 1983.

167. **Eckenrode, C. J., Harman, G. E., and Webb, D. R.,** Seed-borne microorganisms stimulate seedcorn maggot egg laying, *Nature (London),* 256, 487, 1975.

168. **Ehrhardt, P.,** Entwicklung und Symbiontnen geflügelter un ungelflügelter Virgines von *Aphis fabae* Scop. unter dem Einfluss künstlichter Ernährung, *Z. Morphol. Öekol. Tiere,* 57, 295, 1966.

169. **Ehrhardt, P.,** Die Wirkung von Lysozyminjektionen auf Aphiden und deren Symbionten, *Z. Vgl. Physiol.,* 53, 130, 1966.

170. **Ehrhardt, P.,** Einfluss von Ernährungsfaktoren auf die Entwicklung von Säfte saugenden insekten unter besonderer Berücksichtigung von Symbionten, *Z. Parasitenkd.,* 31, 38, 1968.

171. **Ehrhardt, P.,** Die Wirkung verschiedener Spurenelemente auf Wachstum, Reproduktion und Symbionten von *Neomyzus circumflexus* Bukt. (Aphididae, Homoptera, Insecta) bei künstlich Ernährung, *Z. Vgl. Physiol.,* 58, 47, 1968.

172. **Ehrhardt, P.,** Der Vitaminbedarf einer siebröhrensaugenden Aphide, *Neomyzus circumflexus* Buckt., *Z. Vgl. Physiol.,* 60, 416, 1968.

173. **Ehrhardt, P.,** Nachweis durch symbiontische Microorganismen bewirkten Sterinsyntheses in künstlich ernährten Aphiden (Homoptera, Rhynchota, Insecta), *Experientia,* 24, 82, 1968.

174. **Ehrhardt, P. and Schmutterer, H.,** Die Wirkung verschiedener Antibiotica auf Entwicklung und Symbionten Künstlich ernährter

Bohnenblattläuse (*Aphis fabae* Scop.), *Z. Morphol.* Oekol. *Tiere,* 56, 1, 1966.

175. **Ehrhardt, P., Schmutterer, H., and Jayaraj, S.,** Die Wirkung verschiedener, über die Pflanze zugeführter Antibiotica auf Enwicklung and Fertilitt der schwarzen Bohenblattläuse, *Aphis fabae* Scop. *Entomol. Exp. Appl.,* 9, 332, 1966.
176. **Eisenbach, J. and Mittler, T. E.,** Extra-nuclear inheritance in a sexually produced aphid: the ability to overcome host plant resistance by biotype hybrids of the greenbug, *Schizaphis graminum, Experientia,* 43, 332, 1987.
177. **Eisenbach, J. and Mittler, T. E.,** Polymorphism of biotypes E and C of the aphid *Schizaphis graminum* (Homoptera: Aphididae) in response to different scotophases, *Environ. Entomol.,* 16, 519, 1987.
178. **Ellis, P. R., Taylor, J. D., and Littlejohn, I. H.,** The role of microorganisms colonizing radish seedlings in the oviposition behavior of cabbage root fly, *Delia radicum, in Proc. 5th Int. Symp. Insect-Plant Relationships,* Pudoc, Wageningen, The Netherlands, 1982, 297.
179. **van Emden, H. F. and Bashford, M. A.,** A comparison of the reproduction of *Brevicoryne brassicae* and *Myzus persicae* in relations to soluble nitrogen concentration and leaf age (leaf position) in Brussels sprout plant, *Entomol. Exp. Appl.,* 12, 351, 1969.
180. **van Emden, H. F. and Bashford, M. A.,** The performance of *Brevicoryne brassicae* and *Myzus persicae* in relation to plant age and leaf amino acids, *Entomol. Exp. Appl.,* 14, 349, 1971.
181. **van Emden, H. F. and Bashford, M. A.,** The effect of leaf excision on the performance of *Myzus persicae* and *Brevicoryne brassicae* in relation to the nutrient treatments of the plants, *Physiol. Entomol.,* 1, 67, 1976.
182. **Espinoza-Fuentes, F. P. and Terra, W. R.,** Physiological adaptations for digesting bacteria. Water fluxes and distribution of digestive enzymes in *Musca domestica* larval midgut, *Insect Biochem..* 17, 809, 1987.
183. **Eutrick, M. L., O'Brien, R. W., and Slaytor, M.,** Bacteria from the gut of Australian termites, *Appl. Environ. Microbiol.,* 35, 823, 1978.
184. **Eutrick, M. L., Veivers, P., O'Brien, R. W., and Slaytor, M.,** Dependence of the higher termite, *Nasutitermes exitiosus* and the lower termite, *Coptotermes lacteus* on their gut flora, *J. Insect Physiol.,* 24, 363, 1978.
185. **Fine, P. E.,** On the dynamics of symbiote-dependent cytoplasmic incompatibility in culicine mosquitoes, *J. Invertebr. Pathol.,* 30, 10, 1978.
186. **Fink, R.,** Morphologische und physiologische Untersuchungen an den untrazellularen Symbionten von *Pseudococcus citri* Risso, *Z. Morphol. Öekol. Tiere,* 41, 78, 1952.
187. **Fitt, G. P. and O'Brien, R. W.,** Bacteria associated with four species of *Dacus* (Diptera: Tephritidae) and their role in the nutrition of the larvae, *Oecologia,* 67, 447, 1985.
188. **Foeckler,F.,** Reinfektions Versuche steriler Larven von *Stegobium paniceum* L. mit Fremdhefen und die Beziehungen zwischen der Entwicklungsdauer ler Larven und dem Vitamingehalt des Futters und der Hefen, *Z. Morphol. Öekol. Tiere,* 50, 119, 1961.
189. **Foglesong, M. A., Cruden, D. L., and Markovetz, A. J.,** Pleomorphism of *Fusobacteria* isolated from cockroach hindgut, *J. Bacteriol.,* 158, 474, 1984.
190. **Foglesong, M. A., Walker, D. H., Puffer, J. S., and Markovetz, A. J.,** Ultrastructural morphology of some prokaryotic microorganisms associated with the hindgut of cockroaches, *J. Bacteriol.,* 123, 336, 1975.
191. **Forbes, S. A.,** Bacteria normal to digestive organs of Hemiptera, *Bull. Ill. State Lab. Nat. Hist.,* 4, 1, 1892.
192. **Forrest, J. M. S. and Knights, B. A.,** Presence of phytosterols in the food of the aphid, *Myzus persicae, J. Insect Physiol.,* 18, 723, 1972.
193. **Fox, G. E., Stackebrandt, E., Hespell, R. B., Gibson, J., Maniloff, J.,** et al., The phylogeny of prokaryotes, *Science,* 209, 457, 1980.
194. **Francke-Grosmann, H.,** Ectosymbiosus in wood-inhabiting insects, in *Symbiosis,* Vol. 2, Henry, S. M., Ed., Academic Press, New York, 1967, 142.
195. **Fredenhagen, A., Kenny, P., Koji, H., Komura, H., Naya, Y., Nakanishi, K., Nishiyama, K., Sugiura, M., and Tamura, S.,** Role of intracellular symbiotes in planthoppers, in *Pestic. Sci. Biotechnol., Proc. Int. Congr. Pestic. Chem., 6th 1986,* Greenhalgh, R. and Roberts, T. R., Eds., Blackwell Scientific, Oxford, 1987, 101.
196. **French, J. R. J., Turner, G. L., and Bradbury, J. F.,** Nitrogen fixation by bacteria from the hindgut of termites, *J. Gen. Microbiol.,* 95, 202, 1976.
197. **Friend, W. G., Slakeld, E. H., and Stevenson, I. L.,** Nutrition of onion maggots, larvae of *Hylemya antigua* (Meig.), with reference to other members of the genus *Hylemya, Ann. N.Y. Acad. Sci.,* 77, 384, 1959.
198. **Fytizas, E.,** Action de quelques antibiotique sur les adultes de *Dacus oleae* et leur descendance, *Z. Angew. Entomol.,* 65, 453, 1970.
199. **Fytizase, E. and Tzanakakis, M. E.,** Some effects of streptomycin when added to the

adult food, on the adults of *Dacus oleae* and their progeny, *Ann. Entomol. Soc. Am.*, 59, 269, 1966.

200. **Garling, L.**, Origin of ant-fungus mutualism: a new hypothesis, *Biotropica*, 11, 284, 1979.
201. **Gasnier-Fauchet, F. and Nardon, P.**, Comparison of methionine metabolism in symbiotic and aposymbiotic larvae of *Sitphilus oryzae* L. (Coleoptera: Curculionidae). II. Involvement of the symbiotic bacteria in the oxidation of methionine, *Comp. Biochem. Physiol. B*, 85B, 251, 1986.
202. **Gasnier-Fauchet, F. and Nardon, P.**, Comparison of sarcosine and methionine sulfoxide levels in symbiotic and aposymbiotic larvae of two sibling species, *Sitophilus oryzae* L. and *S. zeamais* Mots. (Coleoptera: Curculionidae), *Insect Biochem.*, 17, 17, 1987.
203. **Gasnier-Fauchet, F., Gharib, A., and Nardon, P.**, Comparison of methionine metabolism in symbiotic and aposymbiotic larvae of *Sitophilus oryzae* L. (Coleoptera: Curculionidae). I. Evidence for a glycine N-methyltransferase-like activity in the aposymbiotic larvae, *Comp. Biochem. Physiol. B*, 85B, 245, 1986.
204. **Gassner, G., Duh, F. M., and Bromel, M.**, Chitinolytic activity: a prelude to a symbiotic relationship between bacteria and the screwworm fly?, in *Endocytobiology II, Intracellular Space as Oligogenetic Ecosystem*, Schenk, E. A. and Schwemmier, W., Eds., Walter de Gruyter, Berlin, 1983, 801.
205. **Geigy, R., Halff, L. A., and Kocher, V.**, L'acide folique comme élément important dans la symbiose intestinale de *Triatoma infestans*, *Acta Trop.*, 11, 163, 1954.
206. **Geilman, Q. M.**, The intestinal protozoa of the larvae of the crane fly, *Tipula abdominalis*, *J. Parasitol.*, 19, 173, 1932.
207. **Gier, H. T.**, Intracellular bacteroids in the cockroach, *J. Bacteriol.*, 53, 173, 1947.
208. **Gilmour, D.**, *Biochemistry of Insects*, Academic Press, New York, 1961.
209. **Girolami, V.**, Fruit fly symbiosis and adult survival: general aspects, in *Fruit Flies of Economic Importance*, Cavalloro, R., Ed., A. A. Balkema, Rotterdam, 1983, 74.
210. **Girolami, V.**, Reperti morfo-istologici sulle batteriosimbiosi del *Dacus oleae* Gmelin e di altri ditteri tripetidi, in natura e negli allevamenti su substrati artificiali, Estratto da *REDIA*, 54, 269, 1973.
211. **Glaser, R. W.**, Cultivation and classification of "bacterioids," "symbionts," or "rickettsiae" of *Blatella germanica*, *J. Exp. Med.*, 51, 903, 1930.
212. **Glaser, R. W.**, The intracellular bacteria of the cockroach in relation to symbiosis, *J. Parasitol.*, 32, 483, 1946.
213. **Glasgow, H.**, The gastric caeca and the caecal bacteria of the Heteroptera, *Biol. Bull.*, 26, 101, 1914.
214. **Goeden, R. D. and Norris, D. M.**, Some biological and ecological aspects of ovipositional attack in *Carya* spp. by *Scolytus quadrispinosus* (Coleoptera: Scolytidae), *Ann. Entomol. Soc. Am.*, 58, 771, 1965.
215. **Goodchild, A. J. P.**, The bacteria associated with *Triatoma infestans* and some other species of Reduviidae, *Parasitology*, 45, 441, 1955.
216. **Grassé, P. P.**, Roles des flagellés symbiotiques chez les blattes et les termites, *Tijdschr. Entomol.*, 95, 70, 1952.
217. **Greany, P. D., Tumlinson, J. H., Chambers, D. L., and Boush, G. M.**, Chemically mediated host finding by *Biosteres (Opius) longicaudatus*, a parasitoid of tephritid fruit fly larvae, *J. Chem. Ecol.*, 3, 189, 1977.
218. **Greenberg, B.**, *Flies and Disease, II*, Princeton University Press, Princeton, NJ, 1973, 447.
219. **Gresson, R. A. R. and Threadgold, L. T.**, An electron microscope study of bacteria in the oocytes and follicle cells of *Blatta orientalis*, *Q. J. Microsc. Sci.*, 101, 295, 1960.
220. **Griffiths, G. W. and Beck, S. D.**, Intracellular symbiotes of the pea aphid, *Acyrthosiphon pisum*, *J. Insect Physiol.*, 19, 75, 1973.
221. **Griffiths, G. W. and Beck, S. D.**, Effects of antibiotics on intracellular symbiotes in the pea aphid, *Acyrthosiphon pisum*, *Cell Tissue Res.*, 148, 287, 1974.
222. **Griffiths, G. W. and Beck, S. D.**, Ultrastructure of pea aphid mycetocytes: evidence for symbiote secretion, *Cell Tissue Res.*, 159, 351, 1975.
223. **Griffiths, G. W. and Beck, S. D.**, *In vivo* sterol biosynthesis by pea aphid symbiotes as determined by digitonin and electron microscopic autoradiography, *Cell Tissue Res.*, 176, 179, 1977.
224. **Griffiths, G. W. and Beck, S. D.**, Effect of dietary cholesterol on the pattern of osmium deposition in the symbiote-containing cells of the pea aphid, *Cell Tissue Res.*, 176, 191, 1977.
225. **Grinyer, I. and Musgrave, A. J.**, Microorganisms and mitochondria in the malpighian tubules of *Sitophilus* (Coleoptera), *Can. J. Microbiol.*, 10, 805, 1964.
226. **Grinyer, I. and Musgrave, A. J.**, Ultrastructure and peripheral membranes of the mycetomal microorganisms of *Sitophilus granarius* L. (Coleoptera), *J. Cell. Sci.*, 1, 181, 1966.
227. **Grisham, M. P., Sterling, W. L., Powell, R. D., and Morgan, P. W.**, Characterization of the induction of stress ethylene synthesis in cotton caused by the cotton fleahopper (Hemiptera: Miridae) and its microorga-

nisms, *Ann. Entomol. Soc. Am.*, 80, 411, 1987.

228. **Grosovsky, B. D.-D. and Largulis, L.,** Termite microbial communities, in *Experimental Microbial Ecology,* Burns, R. G. and Slater, J. H., Eds., Blackwell Scientific, Oxford, 1985, 519.
229. **Gupta, M. and Pant, N. C.,** Symbiotes of *Dacus cucurbitae* and their *in vitro* physiology. IV. Function of symbiotes in ovarian development, in *Endocytobiology II, Intracellular Space as Oligogenetic Ecosystem,* Schenk, E. A. and Schwemmier, W., Eds., Walter de Gruyter, Berlin, 1983, 739.
230. **Haas, F. and Koenig, H.,** Characterization of an anaerobic symbiont and the associated aerobic bacterial flora of *Pyrrhocoris apterus* (Heteroptera: Pyrrhocoridae), *FEMS Microbiol. Ecol.,* 45, 99, 1987.
231. **Hagen, K. S.,** Dependence of the olive fruit fly, *Dacus oleae* larvae on the symbiosis with *Pseudomonas savastanoi* for the utilization of the olive, *Nature* (London), 204, 423, 1966.
232. **Hagen, K. S. and Tassan, R. L.,** Exploring nutritional roles of extracellular symbiotes on the reproduction of honeydew feeding adult chrysopids and tephritids, in *Insect and Mite Nutrition,* Rodriguez, J. G., Ed., North-Holland, Amsterdam, 1972, 323.
233. **Hamon, C.,** Etude au microscope électronique des symbiontes à transmission héréditaire chez quelques insectes homoptères-auchénorhynques femelles, *Z. Zellforsch. Mikrosk. Anat.,* 119, 244, 1971.
234. **Hartzell, A.,** Insect ectosymbiosis, in *Symbiosis,* Vol. 1, Henry, S. M., Ed., Academic Press, New York, 1967, 107.
235. **Hasan, R. and Khan, A. M.,** *In vitro* effect of vitamins on the growth of mycetomal symbionts isolated from *Sitophilus granarius* L., *Bull. Pure Appl. Sci. Sect. A,* 3, 30, 1984.
236. **Haydak, M. H. and Dietz, A.,** Cholesterol, panthothenic acid, pyridoxine and thiamine requirements of honey bees for brood rearing, *J. Apic. Res.,* 11, 105, 1972.
237. **Hendry, L. B., Kosteic, J. G., Hindeniang, D. M., Wichmenn, J. K., Fix, C. J., and Korzeniowski, S. H.,** Chemical messengers in insects and plants, *Recent Adv. Phytochem.,* 10, 351, 1976.
238. **Henry, S. M.,** *Symbiosis,* Vol. 2, Academic Press, New York, 1967.
239. **Henry, S. M. and Block, R. J.,** The sulfur metabolism of insects. IV. The conversion of inorganic sulfate to organic sulfur compounds in cockroaches. The role of intracellular symbionts, *Conf. Boyce Thompson Inst. Plant Res.,* 20, 317, 1960.
240. **Hertig, M. and Wolbach, S. B.,** Studies on rickettsia-like microorganisms in insects, *J. Med. Res.,* 44, 329, 1924.
241. **Hervey, A., Rogerson, C. T., and Leong, I.,** Studies on fungi cultivated by ants, *Brittonia,* 29, 226, 1977.
242. **Hill, P., Campbell, J. A., and Petrie, I. A.,** *Rhodnius prolixus* and its symbiotic actinomycete: a microbiological, physiological and behavioural study, *Proc. R. Soc. London Ser. B,* 194, 501, 1976.
243. **Hinde, R.,** Structural and Physiological Studies of the Mycetome Symbiotes of Aphids, Ph.D. Thesis, University of Sydney, Australia, 1970.
244. **Hinde, R.,** Maintenance of aphid cells and the intracellular symbiotes of aphids *in vitro, J. Invertebr. Pathol.,* 17, 333, 1971.
245. **Hinde, R.,** The control of the mycetome symbiotes of the aphids *Brevicoryne brassicae, Myzus persicae* and *Macrosiphum rosae, J. Insect Physiol.,* 17, 1791, 1971.
246. **Hinde, R.,** The fine structure of the mycetome symbiotes of the aphids *Brevicoryne brassicae, Myzus persicae* and *Macrosiphum rosae, J. Insect Physiol.,* 17, 2035, 1971.
247. **Hogan, M. E., Slaytor, M., and O'Brien, R. W.,** Transport of volatile fatty acids across the hindgut of the cockroach *Panesthia cribrata* Saussure and the termite *Mastotermes darwiniensis* Froggatt, *J. Insect Physiol.,* 31, 587, 1985.
248. **Hogan, M. E., Schulz, M. W., Slaytor, M., Czolij, R. T., and O'Brien, R. W.,** Components of termite and protozoal cellulases from the lower termit, *Coptotermes lacteus* Froggatt, *Insect Biochem.,* 18, 45, 1988.
249. **Honigberg, B. M.,** Protozoa associated with termites and their role in digestion, in *Biology of Termites,* Vol. 2, Krishna, K. and Weesner, F. M., Eds., Academic Press, New York, 1970, 1.
250. **Hou, R. F. and Brooks, M. A.,** Continuous rearing of the aster leafhopper, *Macrosteles fascifrons,* on a chemically defined diet, *J. Insect Physiol.,* 21, 1481, 1975.
251. **Hough, J. A., Harman, G. E., and Eckenrode, C. J.,** Microbial stimulation of onion maggot oviposition, *Environ. Entomol.,* 10, 206, 1981.
252. **Houk, E. J.,** Lipids of the primary intracellular symbiote of the pea aphid, *Acyrthosiphon pisum, J. Insect Physiol.,* 20, 471, 1974.
253. **Houk, E. J.,** Maintenance of the primary symbiote of the pea aphid *Acyrthosiphon pisum* in liquid media, *J. Invertebr. Pathol.,* 24, 24, 1974.
254. **Houk, E. J. and Griffiths, G. W.,** Intracellular symbiotes of the Homoptera, *Annu. Rev. Entomol.,* 25, 161, 1980.
255. **Houk, E. J. and McLean, D. L.,** Isolation

of the primary intracellular symbiote of the pea aphid, *Acyrthosiphon pisum, J. Invertebr. Pathol.,* 23, 237, 1974.

256. **Houk, E. J., Griffiths, G. W., and Beck, S. D.,** Lipid metabolism in the symbiotes of the pea aphid, *Acyrthosiphon pisum, Comp. Biochem. Physiol.,* 54B, 427, 1976.
257. **Houk, E. J., McLean, D. L., and Criddle, R. S.,** Pea aphid primary symbiotes deoxyribonucleic acid, *J. Invertebr. Pathol.,* 35, 105, 1980.
258. **Houk, E. J., Griffiths, G. W., Hadjokas, N. E., and Beck, S. D.,** Peptidoglycan in the cell wall of the primary intracellular symbiote of the pea aphid, *Science,* 198, 401, 1977.
259. **Howard, J., Bush, G. L., and Breznak, J. A.,** The evolutionary significance of bacteria associated with *Rhagoletis, Evolution,* 39, 405, 1945.
260. **Howard, R., Matsumura, F., and Coppel, H. C.,** Trail-following pheromones of the Rhinotermitidae: approaches to their authentication and specificity, *J. Chem. Ecol.,* 2, 147, 1976.
261. **Hoyt, C. P., Osborne, G. O., and Mulcock, A. P.,** Production of an insect sex attractant by symbiotic bacteria, *Nature (London),* 230, 472, 1971.
262. **Hungate, R. E.,** Studies on the nutrition of *Zootermopsis.* I. The role of bacteria and molds in cellulose decomposition, *Centr. Bakteriol. II,* 94, 240, 1936.
263. **Hungate, R. E.,** Studies on the nutrition of *Zootermopsis.* II. The relative importance of the termite and protozoa in wood digestion, *Ecology,* 19, 1, 1938.
264. **Hungate, R. E.,** Experiments on the nitrogen economy of termites, *Ann. Entomol. Soc. Am.,* 34, 467, 1941.
265. **Hungate, R. E.,** Quantitative analysis on the cellulose fermentation by termite Protozoa, *Ann. Entomol. Soc. Am.,* 36, 729, 1943.
266. **Iaccarino, F. M. and Tremblay, E.,** Comparazione ultrastrutturale della disimbiosi di *Macrosiphum rosae* (L.) e *Dactynotus jaceae* (L.) (Homoptera, Aphididae), *Boll. Lab. Entomol. Agrar. Portici,* 30, 319, 1973.
267. **Iizuka, T., Horie, Y., and Takizawa, Y.,** The aerobic bacterial flora in the gut of larvae of the silkworm, *Bombyx mori* L. III. Effect of dietary antibiotics on the multiplication of bacteria in the gut and on larval mortality caused by the rearing on artificial diets, *J. Sericult. Sci. Jpn,* 39, 253, 1970.
268. **Inayatullah, C., Webster, J. A., and Fargo, W. S.,** Morphometric variation in the alates of greenbug (Homoptera: Aphididae) biotypes, *Ann. Entomol. Soc. Am.,* 80, 306, 1987.
269. **Ishikawa, H.,** RNA synthesis in aphids, *Lachnus tropicalis, Biochem. Biophys. Res. Commun.,* 78, 1418, 1977.
270. **Ishikawa, H.,** Intracellular symbiont as a major source of the ribosomal RNAs in the aphid mycetocytes, *Biochem. Biophys. Res. Commun.,* 81, 993, 1978.
271. **Ishikawa, H.,** DNA, RNA and protein synthesis in the isolated symbionts from the pea aphid, *Acyrthosiphon pisum, Insect Biochem.,* 12, 605, 1982.
272. **Ishikawa, H.,** Host-symbiont interactions in the protein synthesis in the pea aphid, *Acyrthosiphon pisum, Insect Biochem.,* 12, 613, 1982.
273. **Ishikawa, H.,** Biochemistry of aphid symbionts, in *Endocytobiology,* Vol. 2, Schenk, E. A. and Schwemmier, W., Eds., Walter de Gruyter, Berlin, 1983, 759.
274. **Ishikawa, H.,** Characterization of the protein species synthesized *in vivo* and *in vitro* by an aphid endosymbiont, *Insect Biochem.,* 14, 417, 1984.
275. **Ishikawa, H.,** Age-dependent regulation of protein synthesis in an aphid endosymbiont by the host insect, *Insect Biochem.,* 14, 427, 1984.
276. **Ishikawa, H.,** Control of macromolecule synthesis in the aphid endosymbiont by the host insect, *Comp. Biochem. Physiol. B,* 78, 51, 1984.
277. **Ishikawa, H.,** Molecular aspects of intracellular symbiosis in the aphid mycetocyte, *Zool. Sci.,* 1, 509, 1984.
278. **Ishikawa, H.,** Alteration with age of symbiosis of gene expression in aphid endosymbionts, *BioSystems,* 17, 127, 1984.
279. **Ishikawa, H.,** A species-specific protein of an aphid is produced by its endosymbiont, *Zool. Sci.,* 2, 285, 1985.
280. **Ishikawa, H.,** Symbionin — a protein synthesized by an intracellular symbiotic microorganism in an insect, *Kagaku To Seibutsu,* 23, 346, 1985.
281. **Ishikawa, H.,** The molecular biology of symbiotic bacteria of Aphididae, *Microbiol. Sci.,* 3, 117, 1986.
282. **Ishikawa, H.,** Nucleotide composition and kinetic complexity of the genomic DNA of an intracellular symbiont in the pea aphid *Acyrthosiphon pisum, J. Mol. Evol.,* 24, 205, 1987.
283. **Ishikawa, H. and Hashimoto, H.,** The molecular biology of symbiotic bacteria of Aphididae, *Microbiol. Sci.,* 3, 117, 1986.
284. **Ishikawa, H. and Yamaji, M.,** Protein synthesis by intracellular symbionts in two closely interrelated aphid species, *BioSystems,* 17, 327, 1985.
285. **Ishikawa, H. and Yamaji, M.,** Symbionin, an aphid endosymbiont-specific protein. I. Production of insects deficient in symbiont, *Insect Biochem.,* 15, 155, 1985.
286. **Ishikawa, H., Hashimoto, H., and Yamaji, M.,** Symbionin, an aphid endosymbiont-

specific protein. III. Symbionin present in the male, ovipara and fundatrix, *Insect Biochem.*, 16, 299, 1986.

287. **Ishikawa, H., Yamaji, M., and Hashimoto, H.,** Symbionin, an aphid endosymbiont-specific protein. II. Diminution of symbionin during post-embryonic development of aposymbiotic insects, *Insect Biochem.*, 15, 165, 1985.
288. **Islam, A. and Chatterjee, C.,** On the role of intracellular bacteria in the variation of free amino acids during postembryonic development of *Triolium confusum* (Insecta; Coleoptera; Tenebrionidae), *India J. Zool.*, 12, 47, 1984.
289. **Ito, T. and Tanka, M.,** Rearing of the silkworm by means of aseptic technique, J. *Sericult. Sci. Jpn.*, 31, 7, 1962.
290. **Iverson, K. L., Bromel, M. C., Anderson, A. W., and Freeman, T. P.,** Bacterial symbionts in the sugar beet maggot, *Tetanops myopacformis* (von Röder), *Appl. Environ. Microbiol.*, 47, 22, 1984.
291. **Jayaraj, S., Ehrhardt, P., and Schmutterer, H.,** The effect of certain antibiotics on reproduction of the black bean aphid, *Aphis fabae* Scop., *Ann. Appl. Biol.*, 59, 13, 1967.
292. **Johnson, M. A.,** Biochemistry of conifer resistance to bark beetles and their fungal symbiotes, *ACS Symp. Ser.*, 325, 76, 1987.
293. **Jones, C. G.,** Microorganisms as mediators of plant resource exploitation by insect herbivores, in *A New Ecology: Novel Approaches to Interactive Systems,* Price, P. W., Slobodchikoff, C. N., and Gaud, W. S., Eds., John Wiley & Sons, New York, 1984, 53.
294. **Jones, C. G., Aldrich, J. R., and Blum, M. S.,** 2-Furaldehyde from baldcypress: a chemical rationale for the demise of the Georgia silkworm industry, *J. Chem. Ecol.*, 7, 89, 1981.
295. **Jones, C. G., Aldrich, J. R., and Blum, M. S.,** Baldcypress allelochemics and the inhibition of silkworm enteric microorganisms: some ecological considerations, *J. Chem. Ecol.*, 7, 103, 1981.
296. **Jurzitza, G.,** Der Vitaminbedarf normaler und aposymbiontischer *Lasioderma serricorne* F. (Coleoptera, Anobiidae) und die Bedeutung der symbiontischen Pilze als Vitaminquelle, *Oecologia*, 3, 70, 1969.
297. **Jurzitza, G.,** Die Wirkung des Sulfanilids auf die Symbioten einiger Anobiiden, *Z. Angew. Entomol.*, 52, 302, 1963.
298. **Jurzitza, G.,** Studienan der Symbiose der Anobiiden. II. Physiiologische Studien am Symbioten von *Lasioderma serricorne* F., *Arch. Mikrobiol.*, 49, 331, 1964.
299. **Jurzitza, G.,** The fungi symbiotic with anobiid beetles, in *Insect-Fungus Symbiosis. Nutrition, Mutualism, and Commensalism,* Batra, L. R., Ed., John Wiley & Sons, New York, 1979, 65.
300. **Kellen, W. R. and Hoffman, D. F.,** *Wolbachia* sp. (Rickettsiales: Rickettsiaceae) a symbiont of the almond moth, *Esphestia cautella*: ultrastructure and influence on host fertility, *J. Invertebr. Pathol.*, 37, 273, 1981.
301. **Khan, M. R.,** A study of the digestive system of *Dysdercus fasciatus* Sign. (Pyrrhocoridae, Hemiptera), *Pak. J. Agric. Sci.*, 1, 45, 1964.
302. **Khan, M. R. and Ford, J. B.,** The distribution and localization of digestive enzymes in the alimentary canal and salivary glands of the cotton stainer, *Dysdercus fasciatus, J. Insect Physiol.*, 13, 1619, 1967.
303. **Khokhlacheva, V. E. and Azimdzhanov, I. M.,** Mycoflora of Chinese silkworm caterpillars and mulberry leaves, *Mikol. Fitopatol.*, 11, 248, 1977.
304. **Kimura, M. T.,** Evolution of food preferences in fungus-feeding *Drosophila*: an ecological study, *Evolution,* 34, 1009, 1980.
305. **Kindler, S. D. and Spomer, S. M.,** Biotype status of six greenbug (Homoptera: Aphididae) Isolates, *Environ. Entomol.*, 15, 567, 1986.
306. **Koch, A.,** Uber das Verhalten symbiontenfrier Sitodrepa — Larven, *Biol. Zentralbl.*, 53, 199, 1933.
307. **Koch, A.,** Symbiosestudien. I. Die Symbioses des Splintkäfers, *Lyctus linearis* Goez., *Z. Morphol. Öekol. Tiere,* 32, 137, 1936.
308. **Koch, A.,** Symbiosestudien. II. Experimentelle Untersuchungen an *Oryzaepyhilus surinamensis* L. (Cucujidae, Coleopt.), *Z. Morphol. Öekol. Tiere,* 32, 147, 1936.
309. **Koch, A.,** The experimental elimination of symbionts and its consequences, *Exp. Parasitol.,* 5, 481, 1956.
310. **Koch, A.,** Symbioten-Vitaminquellen der Tiere. Exposé eines Vortrages in der Gesellschaft für Ernahrungsbiologie, E. V., in *Kurzberichte für Landwirtschaft und Ernährung,* Heft 9, W. Brehm, Munich, 1956.
311. **Koch, A.,** Intracellular symbiosis in insects, *Annu. Rev. Microbiol.,* 14, 121, 1960.
312. **Koch, A.,** Insects and their endosymbionts, in *Symbiosis,* Vol. 2, Henry, S. M., Ed., Academic Press, New York, 1967, 1.
313. **Kok, L. T.,** Lipids of ambrosia fungi and the life of mutualistic beetles, in *Insect-Fungus Symbiosis. Nutrition, Mutualism, and Commensalism,* Batra, L. R., Ed., John Wiley & Sons, New York, 1979, 33.
314. **Kolya, A. K. and Pant, N. C.,** Effect of ultraviolet irradiation and centrifugation on the mycetomes and symbiotes of *Oryzaephilus surinamensis, Indian J. Entomol.,* 24, 191,

1962.

315. **Koyama, K.**, Rearing of *Inazuma dorsalis* and *Nephotettix cinticeps* on a synthetic diet, *Jpn. J. Appl. Entomol. Zool.*, 17, 163, 1973.
316. **Koyama, K.**, Rearing of the brown planthopper, *Nilaparavata lugens* Stål (Hemiptera: Delphacidae) on a synthetic diet, *Jpn. J. Appl. Entomol. Zool.*, 23, 39, 1979.
317. **Koyama, K. and Mitsuhashi, J.**, Rearing of the white-backed planthopper, *Sogatella furcifera* Horvath (Hemiptera: Delphacidae), on a synthetic diet, *Jpn. J. Appl. Entomol. Zool.*, 24, 117, 1980.
318. **Körner, H. K.**, Ultrastructure der intrazellulären Symbionten im Embryo der Kleinzikade *Euscelis plebejus* Fall., *Z. Zellforsch. Mikrosk. Anat.*, 100, 466, 1969.
319. **Körner, H. K.**, Die embryonale Entwicklung der symbiontenführenden organe von *Euscekis plebejus* Fall., *Öecologia (Berlin)*, 2, 319, 1969.
320. **Körner, H. K.**, Elektronenmikroskopische Untersuchungen am embryonalen Mycetom der Kleinzikade *Euscelis plebejus* Fall. (Homoptera, Cicadina). I. Die Feinstruktur der a-Symbioten, *Z. Parasitenkd.*, 40, 203, 1972.
321. **Körner, H. K.**, Elektronennmikroskopische Untersuchungen am embryonalen Mycetom der Kleinzikade *Euscelis plebejus* Fall. (Homoptera, Cicadina). II. Die Feinstruktur der t-Symbioten, *Z. Parasitenkd.*, 44, 149, 1974.
322. **Körner, H. K.**, On the host-symbiont cycle of a leafhopper *(Euscelis plebejus)* endosymbiosis, *Experientia*, 32, 463, 1976.
323. **Körner, H. K. and Feldhage, A.**, Inclusions in symbiotic bacteria of a leafhopper (*Euscelis plegejus* Fall.), *Cytobiologie Z. Exp. Zellforsch.*, 1, 203, 1970.
324. **Krieger, D. L.**, Rearing several aphid species on synthetic diet, *Ann. Entomol. Soc. Am.*, 64, 1176, 1971.
325. **Krishna, K. and Weesner, F. M.**, *Biology of Termites*, Vol. 1, Academic Press, New York, 1969.
326. **Krishna, K. and Weesner, F. M.**, *Biology of Termites*, Vol. 2, Academic Press, New York, 1970.
327. **Kukor, J. J. and Martin, M. M.**, Acquisition of digestive enzymes by siricid woodwasps from their fungal symbiont, *Science*, 220, 1161, 1983.
328. **Kukor, J. J. and Martin, M. M.**, Cellulose digestion in *Monochamus marmorator* Kby. (Coleoptera: Cerambycidae): role of acquired fungal enzymes, *J. Chem. Ecol.*, 12, 1057, 1986.
329. **Kukor, J. J. and Martin, M. M.**, The effect of acquired microbial enzymes on assimilation efficiency in the common woodlouse, *Tracheoniscus rathkei*, *Oecologia*, 69, 360, 1986.
330. **Kukor, J. J. and Martin, M. M.**, The transformation of *Saperda calcarata* (Coleoptera, Cerambycidae) into a cellulose digestor through the inclusion of fungal enzymes in its diet, *Oecologia*, 71, 138, 1986.
331. **Kurtii, T. J. and Brooks, M. A.**, Preparation of mycetocytes for culture *in vitro*, *J. Invertebr. Pathol.*, 27, 209, 1976.
332. **Kusumi, T., Suwa, Y., Kita, H., and Nasu, S.**, Symbiotes of planthoppers. I. The isolation of intracellular symbiotes of the smaller brown planthopper *Laodelphax striatellus* Fallén (Hemiptera: Delphacidae), *Appl. Entomol. Zool.*, 14, 459, 1979.
333. **Kusumi, T., Suwa, Y., Kita, H., and Nasu, S.**, Properties of intracellular symbiotes of the smaller brown planthopper *Laodelphax striatellus* Fallén (Hemiptera: Delkphacidae), *Appl. Entomol. Zool.*, 15, 129, 1980.
334. **Lamb, K. P. and Hinde, R.**, Structure and development of mycetome in the cabbage aphid, *Brevicoryne brassicae*, *J. Invertebr. Pathol.*, 9, 3, 1967.
335. **Lane, D. J., Pace, B., Olsen, G. J., Stahl, D. A., Sogin, M. L., and Pace, N. R.**, Rapid determination of 16S ribosomal RNA sequences for phylogenetic analyses, *Proc. Natl. Acad. Sci. U.S.A.*, 82, 6955, 1985.
336. **Lanham, U. N.**, Observations on the supposed intracellular micro-organisms in aphids, *Science*, 115, 459, 1952.
337. **Lanham, U. N.**, The Blochmann bodies: hereditary intra-cellular symbionts in insects, *Biol. Rev.*, 43, 269, 1968.
338. **Laurema, S., Varis, A.-L., and Miettinen, H.**, Studies on enzymes in the salivary glands of *Lygus rugulipennis* (Hemiptera, Miridae), *Insect Biochem.*, 15, 211, 1985.
339. **Le Blanc, N. N. and Musgrave, A. J.**, Some microbiological studies of two species of aphids, *Aphis fabae* Scop. and *Microsiphum pisi* (Harris) (Hemiptera, Homoptera: Aphididae), *Can. J. Microbiol.*, 9, 65, 1963.
340. **Le Fay, A., Courtois, J.-E., Thuillier, A., Chararas, C., and Lambin, S.**, Biochimie-Étude des osidases de l'insecte xylophage *Ips sexdentatus* et de sa flore microbienne, *C. R. Acad. Sci. Paris*, 268, 2968, 1969.
341. **Leckstein, P. M.**, Sulphur-amino acid metabolism in *Aphis fabae*, *Comp. Biochem. Physiol. B*, 49, 743, 1974.
342. **Lederberg, J.**, Cell genetics and hereditary symbiosis, *Physiol. Rev.*, 32, 403, 1952.
343. **Lelung, P. T., Margulls, L., Chase, D., and Nutting, W. L.**, The symbiotic microbial community of the Sonoran desert termite: *Pterotermes occidentis*, *BioSystems*, 13, 109, 1980.
344. **Leufvén, A., Bergström, G., and Falsen, E.**, Interconversion of verbenols and verbenone by identified yeasts isolated from

the spruce bark beetle *Ips typographus, J. Chem. Ecol.,* 10, 1349, 1987.

345. **Lindner, P.,** *Saccharomyces apiculatus parasiticus, Zentralbl. Bakteriol. Abt.,* 2, 1895.
346. **Louis, C.,** Polymorphisme des bactéries symbiotiques du mycétome de *Pseudococcus citri* Risso, *C. R. Acad. Sci. Ser. D,* 260, 1755, 1965.
347. **Louis, C.,** Cytologie et cytochimie du mycétome de *Pseudococcus maritimus* (Ehrhorn), *C. R. Acad. Sci. Ser. D,* 265, 437, 1967.
348. **Louis, C.,** Histochimie et ultrastruture des mycétome et symbiotes d'*Icerya purchasi* Mask., *C. R. Acad. Sci. Ser. D,* 268, 445, 1969.
349. **Louis, C. and Laporte, M.,** Caractères ultrastructuraux et différentiation de formes migratices des symbioties chez *Eucelis plebejus, Ann. Soc. Entomol. Fr.,* 5, 799, 1969.
350. **Louis, C. and Nicolas, G.,** Ultrastructure of the endocellular procaryotes of arthropods as revealed by freeze-etching. I. A study of "a"-type endosymbionts of the leafhopper *Euscelis plebejus, J. Microsc. Biol. Cell.,* 26, 121, 1976.
351. **Louis, C., Nicolas, G., and Pouphile, M.,** Ultrastructure of the endocellular procaryotes of arthropods as revealed by freeze-etching. II. "t"-type endosymbionts of the leafhopper *Euscelis plebejus* Fall., *J. Microsc. Biol. Cell.,* 27, 53, 1976.
352. **Lovelock, M., O'Brien, R. W., and Slaytor, M.,** Effect of laboratory containment on the nitrogen metabolism of termites, *Insect Biochem.,* 15, 503, 1985.
353. **Loxdale, H. D., Rhodes, J. A., and Fox, J. S.,** Electrophoretic study of enzymes from cereal aphid populations. IV. Detection of hidden genetic variation within populations of the grain aphid *Sitobion avenae* (F.) (Hemiptera: Aphididae), *Theor. Appl. Genet.,* 70, 407, 1985.
354. **Lum, P. T. M. and Baker, J. E.,** Development of mycetomes in larvae of *Sitophilus granarius* and *S. oryzae, Ann. Entomol. Soc. Am.,* 66, 1261, 1973.
355. **Madden, J. L. and Couts, M. P.,** The role of fungi n the biology and ecology of woodwasps (Hymenoptera: Siricidae), in *Insect-Fungus Symbiosis. Nutrition, Mutualism, and Commensalism,* Batra, L. R., Ed., John Wiley & Sons, New York, 1979, 165.
356. **Mansour, K.,** Preliminary studies on the bacterial cell-mass (accessory cell-mass) of *Calandra oryzae* (Linn.): the rice weevil, *Q. J. Microsc.,* Sci., 73, 421, 1930.
357. **Mansour, K.,** On the intracellular micro-organisms of some bostrychid beetles, *Q. J. Microsc. Sci.,* 77, 243, 1934.
358. **Mansour, K.,** On the so-called symbiotic relationship between coleopterous insects and intracellular micro-organisms, *Q. J. Micros. Sci.,* 77, 255, 1934.
359. **Mansour, K.,** On the micro-organism-free and the infected *Calandra granaria* (Lin.), *Bull. Soc. Entomol. Egypte,* 19, 290, 1935.
360. **Markkula, M. and Laurema, S.,** The effects of amino acids, vitamins and trace elements on the development of *Acyrthosiphon pisum* (Harris), *Ann. Agric. Fenn.,* 6, 77, 1967.
361. **Marshall, S. A. and Eymann, M.,** Micro-organisms as food for the onion maggot, *Delia (Hylemya) antiqua* (Diptera: Anthomyiidae), *Proc. Entomol. Soc. Ont.,* 112, 1, 1981.
362. **Martin, M. M.,** Biochemical implications of insect mycophagy, *Biol. Rev.,* 54, 1, 1979.
363. **Martin, M. M.,** Cellulose digestion in insects, *Comp. Biochem. Physiol.,* 75A, 313, 1983.
364. **Martin, M. M.,** The role of ingested enzymes in the digestive processes of insects, in *Animal-Microbial Interactions,* Anderson, J. M., Rayner, A. D. M., and Walton, D., Eds., Cambridge University Press, Cambridge, 1984, 155.
365. **Martin, M. M. and Kukor, J. J.,** Role of mycophagy and bacteriophagy in invertebrate nutrition, in *Current Perspectives in Microbial Ecology,* Klug, M. J. and Reddy, C. A., Eds., American Society for Microbiology, Washington, D.C., 1984, 257.
366. **Martin, M. M. and Martin, J. S.,** Cellulose digestion in the midgut of the fungus growing termite, *Macrotermes natalensis:* the role of acquired digestive enzymes, *Science,* 199, 1453, 1978.
367. **Martin, M. M. and Martin, J. S.,** The distribution and origins of the cellulolytic enzymes of the higher termite, *Macrotermes natalensis, Physiol. Zool.,* 52, 11, 1979.
368. **Martin, M. M., Gieselmann, M. J., and Martin, J. S.,** Rectal enzymes of attine ants. a-Amylase and chitinase, *J. Insect Pathol.,* 19, 1409, 1973.
369. **Martin, M. M., Boyd, N. D., Gieselmann, M. J., and Silver, R. G.,** Activity of faecal fluid of a leaf-cutting ant toward plant cell wall polysaccharides, *J. Insect Physiol.,* 21, 1887, 1975.
370. **Martin, M. M., Kukor, J. J., Martin, J. S., and Merritt, R. W.,** The digestive enzymes of larvae of the black fly, *Prosimulium fuscum* (Diptera, Simuliidae), *Comp. Biochem. Physiol.,* 82B, 37, 1985.
371. **Martin, M. M., Martin, J. S., Kukor, J. J., and Merritt, R. W.,** The digestion of protein and carbohydrate by the stream detritivore, *Tipula abdominalis* (Diptera, Tipulidae), *Oecologia,* 46, 360, 1980.
372. **Martin, M. M., Martin, J. S., Kukor, J. J., and Merritt, R. W.,** The digestive enzymes of detritus-feeding stonefly nymphs (Ple-

coptera: Pteronarcyidae), *Can. J. Zool.*, 59, 1947, 1981.

373. **Martin, M. M., Kukor, J. J., Martin, J. S., Lawson, D. L., and Merritt, R. W.**, Digestive enzymes of larvae of three species of caddisflies (Trichoptera), *Insect Biochem.*, 11, 501, 1981.
374. **Martin, M. M., Kukor, J. J., Martin, J. S., O'Toole, T. E., and Johnson, M. W.**, Digestive enzymes of fungus-feeding beetles, *Physiol. Zool.*, 54, 137, 1981.
375. **Martin, W. R., Jr., Grisham, M. P., Kenerley, C. M., Sterling, W. L., and Morgan, P. W.**, Microorganisms associated with cotton fleahooper, *Pseudatomoscelis seriatus* (Heteroptera: Miridae), *Ann. Entomol. Soc. Am.*, 80, 251, 1987.
376. **Martin, W. R., Jr., Morgan, P. W., Sterling, W. L., and Kenerley, C. M.**, Cotton fleahopper and associated microorganisms as components in the production of stress ethylene by cotton, *Plant Physiol.*, in press.
377. **Martin, W. R., Jr., Morgan, P. W., Sterling, W. L., and Meola, R. W.**, Stimulation of ethylene production in cotton by salivary enzymes of the cotton fleahopper, *Pseadatomoscelis seriatus* (Heteroptera: Miridae), *Environ. Entomol.*, in press.
378. **Martin, W. R., Jr., Sterling, W. L., Kenerley, C. M., and Morgan, P. W.**, Transmission of bacterial blight of cotton, *Xanthomonas campestris* pv. *Malvacearum*, by feeding of the cotton fleahopper: implications for stress ethylene-induced square loss in cotton, *J. Entomol. Sci.*, 23, 161, 1988.
379. **Matha, V., Soldan, T., and Weyda, F.**, In vitro synthesis of DNA and proteins in mycetomes of the tsetse fly, *Glossina palpalis palpalis* (Diptera, Glossinidae), *Microbios Lett.*, 30, 147, 1985.
380. **Matsubara, F., Kato, M., Hayashiya, K., Kodama, R., and Hanamura, Y.**, Aseptic rearing of silkworm with prepared food, *J. Sericult. Sci. Jpn.*, 36, 39, 1967.
381. **Matsumura, F. and Benezet, H. J.**, Microbial degradation of insecticides, in *Pesticide. Microbiology,* Hill, I. R. and Wright, S. J. L., Eds., Academic Press, London, 1978, 623.
382. **Matsumura, F., Coppel, H. C., and Tal, A.**, Isolation and identification of termite trail following pheromone, *Nature (London),* 219, 963, 1968.
383. **Matsumura, F., Nishimoto, K., Ikeda, T., and Coppel, H. C.**, Influence of carbon sources on the production of the termite trail-following substance by *Gloeophyllum trabeum, J. Chem. Ecol.*, 2, 299, 1976.
384. **Mauldin, J. K.**, Cellulase catabolism and lipid synthesis by normally and abnormally faunated termites, *Reticulitermes flavipes, Insect Biochem.*, 7, 27, 1977.
385. **Mauldin, J. K., Rich, N. M., and Cook, D. W.**, Amino acid synthesis from ^{14}C-acetate by normally and abnormally faunated termites, *Coptermes formosanus, Insect Biochem.*, 8, 105, 1978.
386. **Mayo, Z. B., Starks, K. J., Banks, D. J., and Veal, R. A.**, Variation in chromosome length among five biotypes of the greenbug (Homoptera: Aphididae), *Ann. Entomol. Soc. Am.*, 81, 128, 1988.
387. **McEwen, S. E., Slaytor, M., and O'Brien, R. W.**, Cellobiase activity in three species of Australian termites, *Insect Biochem.*, 10, 563, 1980.
388. **McLean, D. L. and Houk, E. J.**, Phase contrast and electron microscopy of the mycetocytes and symbiotes of the pea aphid, *Acyrthosiphon pisum, J. Insect Physiol.*, 19, 625, 1973.
389. **Meek, S. R.**, Occurrence of *Rickettsia*-like symbionts among species of the *Aedes scutellaris* group (Diptera: Culicidae), *Ann. Trop. Med. Parasitol.*, 78, 377, 1984.
390. **Miles, P. W.**, The saliva of Hemiptera, *Adv. Insect Physiol.*, 9, 183, 1972.
391. **Miles, P. W.**, Plant-sucking bugs can remove the contents of cells without mechanical damage, *Experientia,* 43, 937, 1987.
392. **Milne, D. L.**, The mechanisms of growth retardation by nicotine in the cigarette beetle, *Lasioderma serricorne, S. Afr. J. Agric. Sci.*, 4, 277, 1961.
393. **Milne, D. L.**, A study of the nutrition of the cigarette beetle, *Lasioderma serricorne* F. (Coleoptera: Anobiidae) and a suggested new method for its control, *J. Entomol. Soc. S. Afr.*, 26, 43, 1963.
394. **Mitsuhashi, J.**, Cultivation of intracellular yeast-like organisms in the smaller brown planthopper, *Laodelphax striatellus* Fallén (Hemiptera, Delphacidae), *Appl. Entomol. Zool.*, 10, 243, 1975.
395. **Mitsuhashi, J. and Kono, Y.**, Intracellular microorganisms in the green rice leafhopper, *Nephotettix cincticeps* Uhler (Hemiptera: Deltocephalidae), *Appl. Entomol. Zool.*, 10, 1, 1975.
396. **Mitsuhashi, J. and Koyama, K.**, Rearing of planthoppers on a holidic diet, *Entomol. Exp. Appl.*, 14, 93, 1971.
397. **Mitsuhashi, J. and Koyama, K.**, Artificial rearing of the smaller brown planthopper, *Laodelphax striatellus* Fallén, with special reference to rearing conditions for the first instar nymphs, *Jpn. J. Appl. Entomol. Zool.*, 16, 8, 1972.
398. **Mittler, T. E.**, Amino-acids in phloem sap and their excretion by aphids, *Nature (London),* 172, 207, 1953.

399. **Mittler, T. E.,** Studies on the feeding and nutrition of *Tuberlachnus salignus.* III. The nitrogen economy, *J. Exp. Biol.,* 35, 626, 1958.
400. **Mittler, T. E.,** Effect of amino acid and sugar concentrations on the food uptake of the aphid *Myzus persicae, Entomol. Exp. Appl.,* 10, 39, 1967.
401. **Mittler, T. E.,** Gustations of dietary amino acids by the aphid *Myzus persicae, Entomol. Exp. Appl.,* 10, 87, 1967.
402. **Mittler, T. E.,** Dietary requirements of the aphid *Myzus persicae* affected by antibiotic uptake, *J. Nutr.,* 101, 1023, 1971.
403. **Mittler, T. E.,** Some effects on the aphid *Myzus persicae* of ingesting antibiotics incorporated into artificial diets, *J. Insect Physiol.,* 17, 1333, 1971.
404. **Mittler, T. E.,** Application of artificial feeding techniques for aphids, in *Aphids: Their Biology, Natural Enemies and Control,* Minks, A. K. and Harrewijn, P., Eds., Elsevier, Amsterdam, in press.
405. **Mittler, T. E. and Dadd, R. H.,** Artificial feeding and rearing of the aphid, *Myzus persicae* (Suizer), on a completely defined synthetic diet, *Nature (London),* 195, 404, 1962.
406. **Mittler, T. E. and Koski, P.,** Meridic artificial diets for rearing aphids, *Entomol. Exp. Appl.,* 17, 524, 1974.
407. **Miyazaki, S., Boush, G. M., and Baerwald, R. J.,** Amino acid synthesis by *Pseudomonas melophthora,* bacterial symbiote of *Rhagoletis pomonella* (Diptera), *J. Insect Physiol.,* 14, 513, 1968.
408. **Moeck, H. A.,** Ethanol as the primary attractant for the ambrosia beetle *Trypodendron lineatum* (Coleoptera: Scolytidae), *Can. Entomol.,* 102, 985, 1970.
409. **Montllor, C. B.,** Aphid feeding behavior and performance in relation to plant chemicals, in *Focus on Insect-Plant Interactions,* Bernays, E. A., Ed., CRC Press, Boca Raton, FL, in preparation.
410. **Moore, B. P.,** Biochemical studies in termites, in *Biology of Termites,* Vol. 1, Krishna, K. and Weesner, F. M., Eds., Academic Press, New York, 1969, 407.
411. **Morgan, M. R. J.,** Gut carbohydrases in locusts and grasshoppers, *Acrida,* 5, 45, 1976.
412. **Mullins, D. E. and Cochran, D. G.,** Nitrogen metabolism in the American cockroach. I. An examination of positive nitrogen balance with respect to mobilization of uric acid stores, *Comp. Biochem. Physiol.,* 50A, 489, 1975.
413. **Mullins, D. E. and Cochran, D. G.,** Nitrogen in the American cockroach. II. An examination of negative nitrogen balance with respect to mobilization of uric acid stores, *Comp. Biochem. Physiol.,* 50A, 501, 1975.
414. **Musgrave, A. J.,** Insect mycetomes, *Can. Entomol.,* 96, 377, 1964.
415. **Musgrave, A. J. and Grinyer, I.,** Membranes associated with the disintegration of mycetomal microorganisms in *Sitophilus zeamais* (Mots.) (Coleoptera), *J. Cell. Sci.,* 3, 65, 1968.
416. **Musgrave, A. J. and Homan, R.,** *Sitophilus sasakii* (Tak.) (Coleoptera: Curculionidae) in Canada: anatomy and mycetomal symbiotes as valid taxonomic characters, *Can. Entomol.,* 94, 1196, 1962.
417. **Musgrave, A. J. and McDermott, L. A.,** Some media used in an attempt to isolate and culture the mycetomal microorganisms of *Sitophilus* weevils, *Can. J. Microbiol.,* 7, 842, 1961.
418. **Musgrave, A. J. and Miller, J. J.,** Some microorganisms associated with the weevils *Sitophilus granarius* L. and *Sitophilus oryza* L. I. Distribution and description of the organisms, *Can. Entomol.,* 85, 387, 1953.
419. **Musgrave, A. J. and Miller, J. J.,** Some micro-organisms associated with the weevils *Sitophilus granarius* L. and *Sitophilus oryza* L. II. Population differences of mycetomal micro-organisms in different strains of *S. granarius, Can. Entomol.,* 88, 97, 1956.
420. **Musgrave, A. J. and Miller, J. J.,** Studies of the association between strains and species of *Sitophilus* weevils and their mycetomal micro-organisms, *Proc. 10th Int. Congr. Entomol. Montreal,* 2, 315, 1958.
421. **Musgrave, A. J. and Singh, S. B.,** Histochemical evidence of nuclear equivalents in mycetomal microorganisms in *Sitophilus granarius* (Linnaeus), *J. Invertebr. Pathol.,* 7, 269, 1965.
422. **Musgrave, A. J., Ashton, G. C., and Homan, R.,** Quantitative and qualitative effects of temperature and type of grain on populations of *Sitophilus* (Coleoptera: Curculionidae) and on their mycetomal microorganisms, *Can. J. Zool.,* 41, 1245, 1963,.
423. **Musgrave, A. J., Grinyer, I., and Homan, R.,** Some aspects of the fine structure of the mycetomes and mycetomal micro-organisms in *Sitophilus* (Coleoptera: Curculionidae). *Can. J. Microbiol.,* 8, 747, 1962.
424. **Musgrave, A. J., Homan, R., and Grinyer, I.,** Mycetomal and other microorganisms in young and aging *Sitophilus* (Coleoptera: Curculionidae), *Can. J. Microbiol.,* 10, 806, 1964.
425. **Musgrave, A. J., Monro, H. A. U., and Upitis, E.,** Apparent effect on the mycetomal micro-organisms of repeated exposure of the host insect, *Sitophilus granarius* L. (Coleoptera) to methyl bromide fumigation, *Can. J. Microbiol.,* 7, 280, 1961.
426. **Nakasuji, Y. and Kodama, R.,** Bacteria iso-

lated from silkworm larvae. V. Identification of gram-negative bacteria and their pathogenic effects on asceptically reared silkworm larvae, *J. Sericult. Sci. Jpn.*, 38, 471, 1969.

427. **Nardon, P.**, Contribution à l'étude des symbiotes ovariens de *Sitophilus sasakii*. localisation, histochimie et ultrastructure chez la femelle adulte, *C. R. Acad. Sci. Ser. D*, 272, 2975, 1971.
428. **Nardon, P.**, Obtention d'une souche asymbiotique chez le charancon *Sitophilus sasakii* Tak.: différentes méthodes et comparaison avec la souche symbiotique d'origine, *C. R. Acad. Sci. Ser. D*, 277, 981, 1973.
429. **Nasu, S., Kusumi, T., Suwa, Y., and Kita, H.**, Symbiotes of planthoppers. II. Isolation of intracellular symbiotic microorganisms from the brown planthopper, *Nilaparvata lugens* Stål, and immunological comparison of the symbiotes associated with rice planthoppers (Hemiptera: Delphacidae), *Appl. Entomol. Zool.*, 16, 88, 1981.
430. **Nazarczuk, R. A., O'Brien, R. W., and Slaytor, M.**, Alteration of gut microbiota and its effect on nitrogen metabolism in termites, *Insect Biochem.*, 11, 267, 1981.
431. **Nes, W. D., Campbell, B. C., and Stafford, A. E.**, Metabolism of mevalonic acid to long chain fatty alcohols in an insect, *Biochem. Biophys. Res. Commun.*, 108, 1258, 1982.
432. **Nes, W. R. and Nes, W. D.**, *Lipids in Evolution*, Plenum Press, New York, 1980, 244.
433. **Niemierko, S.**, Some aspects of lipid metabolism in insects, *Proc. 4th Int. Congr. Biochem.*, 12, 185, 1959.
434. **Noda, H.**, Histological and histochemical observation of intracellular yeastlike symbiotes in the fat body of the smaller brown planthopper, *Laodelphax striatellus* (Homoptera: Delphacidae), *Appl. Entomol. Zool.*, 12, 134, 1977.
435. **Noda, H. and Mittler, T. E.**, Sterol biosynthesis by symbiotes of aphids and leafhoppers, in *Metabolic Aspects of Lipid Nutrition in Insects*, Mittler, T. E. and Dadd, R. H., Eds., Westview Press, Boulder, CO, 1983, 41.
436. **Noda, H. and Saito, T.**, Effects of high temperature on the development of *Laodelphax striatellus* (Homoptera: Delphacidae) and on its intracellular yeastlike symbiotes, *Appl. Entomol. Zool.*, 14, 64, 1979.
437. **Noda, H. and Saito, T.**, The role of intracellular yeastlike symbiotes in the development of *Laodelphax striatellus* (Homoptera: Delphacidae), *Appl. Entomol. Zool.*, 14, 453, 1979.
438. **Noda, H., Wada, K., and Saito, T.**, Sterols in *Laodelphax striatellus* with special reference to the intracellular yeastlike symbiotes as a sterol source, *J. Insect Physiol.*, 25, 443, 1979.
439. **Nogge, G.**, Aposymbiotic tsetse flies, *Glossina morsitans morsitans* obtained by feeding on rabbits immunized specifically with symbionts, *J. Insect Physiol.*, 24, 299, 1978.
440. **Nogge, G.**, Elimination of symbionts of tsetse flies (*Glossina m. morsitans* Westwood) by help of specific antibodies, in *Endocytobiology*, Schwemmier, W. and Schenk, H. E. A., Eds., Walter de Gruyter, Berlin, 1980, 445.
441. **Nogge, G. and Gerresheim, A.**, Experiments on the elimination of symbionts from the tsetse fly, *Glossina morsitans morsitans* (Diptera: Glossinidae), by antibiotics and lysozyme, *J. Invertebr. Pathol.*, 40, 166, 1982.
442. **O'Brien, R. W. and Slaytor, M.**, Role of microorganisms in the metabolism of termites, *Aust. J. Biol. Sci.*, 35, 239, 1982.
443. **Odelson, D. A. and Breznak, J. A.**, Volatile fatty acid production by the hindgut microbiota of xylophagous termites, *Appl. Environ. Microbiol.*, 45, 1602, 1983.
444. **Olsen, G. J.**, Comparative Analysis of Nucleotide Sequence Data, Ph.D. thesis, University of Colorado, Denver, 1983, 163.
445. **Olsen, G. J., Lane, D. J., Giovannoni, S. J., and Pace, N. R.**, Microbial ecology and evolution: a ribosomal RNA approach, *Annu. Rev. Microbiol.*, 40, 337, 1986.
446. **Orenski, S. W., Mitsuhashi, J., Ringel, S. M., Martin, J. F., and Maramorosch, K.**, A presumptive bacterial symbiont from the eggs of the six-spotted leafhopper, *Macrosteles fascifrons* Stål., *Contrib. Boyce Thompson Inst.*, 23, 123, 1965.
447. **Pant, N. C. and Dang, K.**, Physiology and elimination of intracellular symbiotes in some stored product beetles, in *Insect and Mite Nutrition*, Rodriguez, J. G., Ed., North-Holland, Amsterdam, 1972, 311.
448. **Pant, N. C. and Fraenkel, G.**, The function of the symbiotic yeasts of two insect species, *Lasioderma serricorne* F. and *Stegobium (Sitodrepa) paniceum* L., *Science*, 112, 498, 1950.
449. **Pant, N. C. and Fraenkel, G.**, On the function of the intracellular symbionts of *Oryzaephilus surinamensis* L. (Cucujidae: Coleoptera), *J. Zool. Soc. India*, 6, 173, 1954.
450. **Pant, N. C. and Fraenkel, G.**, Studies on the symbiotic yeasts of two insect species, *Lasioderma serricorne* F. and *Stegobium paniceum* L., *Biol. Bull.*, 107, 420, 1954.
451. **Pasti, M. B. and Belli, M. L.**, Cellulolytic activity of actinomycetes isolated from termites (Termitidae) gut, *FEMS*, 26, 83, 1985.
452. **Petri, L.**, Recherche sopra i batteri intestinali della mosca olearea, *Mem. Staz. Pat. Veg., Roma*, 1909, 129.
453. **Petri, L.**, Untersuchungen über die Darm-

Bakterien der Olivefliege, *Zentralbl. Bakteriol. Parasitenkd. Infektionskr. Abt. 2,* 26, 357, 1910.

454. **Piekarski, G.,** Electron microscopy of lice symbionts, in *Endocytobiology,* Schwemmler, W. and Schenk, H. E. A., Eds., Walter de Gruyter, Berlin, 1980, 417.

455. **Pitman, G. B., Hedden, R. L., and Gara, R. I.,** Synergistic effects of ethyl alcohol on the aggregation of *Dendroctonus pseudotsugae* (Col., Scolytidae) in response to pheromones, *Z. Angew. Entomol.,* 78, 203, 1975.

456. **Poinar, G. O., Jr., Thomas, G., and Prokopy, R. J.,** Microorganisms associated with *Rhagoletis pomonella* (Tephritidae: Diptera) in Massachusetts, *Proc. Entomol. Soc. Am.,* 108, 19, 1972.

457. **Pollard, D. G.,** Plant penetration by feeding aphids (Hemiptera, Aphidoidea): a review, *Bull. Entomol. Res.,* 62, 631, 1973.

458. **Ponsen, M. B.,** Anatomy of an aphid vector: *Myzus persicae* in *Aphids as Virus Vectors,* Harris, K. F. and Maramorosch, K., Eds., Academic Press, New York, 1976, 63.

459. **Portier, P.,** Passage de l'asepsie a l'envahissement symbiotique humoral et tissulaire par les microorganismes dans la serie des larves des insectes, *C. R. Séances Soc. Biol.,* 70, 916, 1911.

460. **Potrikus, C. J. and Breznak, J. A.,** Nitrogen-fixing *Enterobacter agglomerans* isolated from guts of wood-eating termites, *Appl. Environ. Microbiol.,* 33, 392, 1977.

461. **Potrikus, C. J. and Breznak, J. A.,** Uric acid degrading bacteria in guts of termites [Reticulitermes flavipes (Kollar)], *Appl. Environ. Microbiol.,* 40, 117, 1980.

462. **Potrikus, C. J. and Breznak, J. A.,** Anaerobic degradation of uric acid by gut bacteria of termites, *Appl. Environ. Microbiol.,* 40, 125, 1980.

463. **Potrikus, C. J. and Breznak, J. A.,** Uric acid in wood-eating termites, *Insect Biochem.,* 10, 19, 1980.

464. **Potrikus, C. J. and Breznak, J. A.,** Gut bacteria recycle uric acid nitrogen in termites: a strategy for nutrient conservation, *Proc. Natl. Acad. Sci. U.S.A.,* 78, 4601, 1981.

465. **Potts, R. C. and Hewitt, P. H.,** The partial purification and some properties of the cellulase from the termite *Trinervitermes trinervoides* (Nasutitermitinae), *Comp. Biochem. Physiol.,* 47B, 317, 1974.

466. **Potts, R. C. and Hewitt, P. H.,** Some properties and reaction characteristics of the partially purified cellulase from the termite *Trinervitermes trinervoides* (Nasutitermitinae), *Comp. Biochem. Physiol.,* 47B, 327, 1974.

467. **Powell, R. J. and Stradling, D. J.,** Factors influencing the growth of *Attamyces bromatificus,* a symbiont of attine ants, *Trans. Br. Mycol. Soc.,* 87, 205, 1986.

468. **Prestwich, G. D.,** Defense mechanisms of termites, *Annu. Rev. Entomol.,* 29, 201, 1984.

469. **Prestwich, G. D. and Bently, B. L.,** Nitrogen fixation by intact colonies of the termite *Nasutitermes corniger, Oecologia,* 49, 249, 1981.

470. **Prestwich, G. D., Bently, B. L., and Carpenter, E. J.,** Nitrogen sources for neotropical nasute termites: fixation and selective foraging, *Oecologia,* 46, 297, 1980.

471. **Puchta, O.,** Experimentelle Untersuchungen über die Bedeutung der Symbiose der Kleiderlaus *Pediculus vestimenti* Burm., *Z. Parasitenkd.,* 17, 1, 1955.

472. **Purcell, A. H., Steiner, T., Megraud, F., and Bove, J.,** *In vitro* isolation of a transovarially transmitted bacterium from the leafhopper *Euscelidius variegatus* (Hemiptera: Cicadellidae), *J. Invertebr. Pathol.,* 48, 66, 1986.

473. **Raffa, K. F. and Berryman, A. A.,** Accumulation of monoterpenes and associated volatiles following inoculation of grand fir with a fungus transmitted by the fir engraver, *Scolytus ventralis* (Coleoptera: Scolytidae), *Can. Entomol.,* 114, 797, 1982.

474. **Raffa, K. F. and Berryman, A. A.,** Physiological differences between lodgepole pines resistant and susceptible to the mountain pine beetle and associated microorganisms, *Environ. Entomol.,* 11, 486, 1982.

475. **Ragsdale, D. W., Larson, A. D., and Newsom, L. D.,** Microorganisms associated with feeding and from various organs of *Nezara viridula, J. Econ. Entomol.,* 72, 725, 1979.

476. **Rao, K. D. P., Norris, D. M., and Chu, H. M.,** Lipid interdependencies between *Xyloborus* ambrosia beetles and their ectosymbiotic microbes, in *Metabolic Aspects of Lipid Nutrition in Insects,* Mittler, T. E. and Dadd, R. H., Eds., Westview Press, Boulder, CO, 1983, 27.

477. **Rasmussen, R. A. and Khalil, M. A. K.,** Global production of methane by termites, *Nature (London),* 301, 700, 1983.

478. **Ratner, S. S.,** Structure and Function of the Esophageal Bulb of the Apple Maggot Fly, *Rhagoletis pomonella* Walsh, Ph.D. thesis, University of Massachusetts, Amherst, 1981.

479. **Ratner, S. S. and Stoffolano, J. G.,** Development of the oesophageal bulb of the apple maggot, *Rhagoletis pomonella* (Diptera: Tephritidae): morphological, histological, and histochemical study, *Ann. Entomol. Soc. Am.,* 75, 555, 1982.

480. **Reinhardt, C., Steiger, R., and Hecker, H.,** Ultrastructural study of the midgut mycetome-bacteroids of the tsetse flies *Glossina morsitans, G. fuscipes* and *G. brevipalpis, Acta Trop.,* 280, 1972.

481. **Reis, E.**, Die Symbiose der Lause und Federlinge, *Z. Morphol. Öekol. Tiere,* 20, 233, 1931.

482. **Renwick, J. A. A., Vité, J. P., and Billings, R. F.**, Aggregation pheromones in the ambrosia beetle *Platypus flavicornis, Naturwissenschaften,* 64, 226, 1977.

483. **Retnakaran, A. and Beck, S. D.**, Amino acid requirements and sulfur amino acid metabolism in the pea aphid, *Acyrthosiphon pisum* (Harris), *Comp. Biochem. Physiol.,* 24, 611, 1968.

485. **Richards, A. G. and Brooks, M. A.**, Internal symbioses in insects, *Annu. Rev. Entomol.,* 3, 37, 1958.

486. **Ritter, K. S., Weiss, B. A., Norribom, A. L., and Nes, W. R.**, Identification of $D^{5,7}$-24-methylene and methyl sterols in the brain and whole body of *Atta cephaiotes isthmicola, Comp. Biochem. Physiol.,* 71B, 345, 1982.

487. **Rochaix, J. D.**, Cyclization of chloroplast DNA fragments of *Chlamydomonas reinhardi, Nature (London) New Biol.,* 238, 76, 1972.

488. **Rossiter, M. C., Howard, D. J., and Bush, G. L.**, Symbiotic bacteria of *Rhagoletis pomonella,* in *Fruit Flies of Economic Importance,* Cavalloro, R., Ed., A. A. Balkema, Rotterdam, 1983, 77.

489. **Sacchi, L., Grigolo, A., and Laudani, U.**, Behavior of symbionts during oogenesis and early stages of development in the German cockroach, *Blatella germanica* (Blaitoidea), *J. Invertebr. Pathol.,* 46, 139, 1985.

490. **Sander, K.**, Entwicklungsphysiologische Untersuchungen am embryonalen Mycetom von *Euscelis plebejus* Fall. I. Ausschaltung und abnorme Kombination einzeiner Komponenten des symbiontischen Systems, *Dev. Biol.,* 17, 16, 1968.

491. **Sands, W. A.**, Association of termites and fungi, in *Biology of Termites,* Vol. 1, Krishna, K. and Weesner, F. M., Eds., Academic Press, New York, 1969, 495.

492. **Sanger, R., Nicklen, S., and Coulson, A. R.**, DNA sequencing with chain-terminating inhibitors, *Proc. Natl. Acad. Sci. U.S.A.,* 74, 5463, 1977.

493. **Scheinert, W.**, Symbiose und Embryonalentwicklung bei Rüsselkäfern, *Z. Morphol. Öekol. Tiere,* 27, 76, 1933.

494. **Schneider, H.**, Morphologische und experimentelle Untersuchungen über die Endosymbiose der Korn- und Reiskäfer (*Calandra granaria* L. und *Calandra oryzae* L.), *Z. Morphol. Öekol. Tiere,* 44, 555, 1956.

495. **Schneider, W. D., Miller, J. R., Breznak, J. A., and Fobes, J. F.**, Onion maggot, *Delia antiqua,* survival and development on onions in the presence and absence of microorganisms, *Entomol. Exp. Appl.,* 33, 50, 1983.

496. **Schultz, J. E. and Breznak, J. A.**, Heterotrophic bacteria present in hindguts of wood-eating termites [*Reticulitermes flavipes* (Kollar)], *Appl. Environ. Microbiol.,* 35, 930, 1978.

497. **Schulz, J. E. and Breznak, J. A.**, Cross-feeding of lactate between *Streptococcus lactis* and *Bacteroides* sp. isolated from termite hindguts, *Appl. Environ. Microbiol.,* 37, 1206, 1979.

498. **Schwemmler, W.**, Sprengung der Endosymbiose von *Euscelis plebejus* F. und Ernährung aposymbiontischer Tiere mit synthetischer Diät (Hemiptera, Cicadidae), *Z. Morphol. Tiere,* 74, 297, 1973.

499. **Schwemmler, W.**, Beitag zur Analyse des Endosymbiosezyklus von *Euscelis plebejus* F. Mittels *in vitro* beobachtung, *Biol. Zentralbl.,* 92, 749, 1973.

500. **Schwemmler, W.**, *In vitro* Vermehrung intrazellulärer Zikaden-Symbioten und Reinfektion asymbiontishcer Mycetocyten-Kulturen, *Cytobios,* 8, 63, 1973.

501. **Schwemmler, W.**, Zikaden leben mit dem Erbgut ihrer Symbionten, *Umsch. Wiss. Tech.,* 73, 438, 1973.

502. **Schwemmler, W.**, Studies on the fine structure of leafhopper intracellular symbionts during their reproductive cycles, *Appl. Entomol. Zool.,* 9, 215, 1974.

503. **Schwemmler, W.**, Zikadenendosymbiose: ein Modell für die Evolution höherer Zellen, *Acta Biotheor.,* 23, 132, 1974.

504. **Schwemmler, W.**, Control mechanisms of leafhopper endosymbiosis, in *Contemporary Topics in Immunology,* Vol. 4, Cooper, E. L., Ed., Plenum Press, New York, 1974, 179.

505. **Schwemmler, W., Quiot, J.-M., and Amargier, A.**, Étude de la symbiose intracellulair sur cultures organotypiques et cellulaire de l'homoptère, *Euscelis plebejus* en milieux a fractions standard, *Ann. Soc. Entomol. Fr.,* 7, 423, 1971.

506. **Sharma, B. R., Martin, M. M., and Shafer, J. A.**, Alkaline proteases from the gut fluids of detritus-feeding larvae of the crane fly, *Tipula abdominalis* (Say) (Diptera, Tipulidae), *Insect Biochem.,* 14, 37, 1984.

507. **Simon, J.-P., Parent, M.-A., and Auclair, J.-L.**, Isozyme analysis of biotypes and field populations of the pea aphid, *Acyrthosiphon pisum, Entomol. Exp. Appl.,* 32, 186, 1982.

508. **Sigh, S. B. and Musgrave, A. J.**, Some studies on the chromatin and cell wall of the mycetomal microorganisms of *Sitophilus granarius* L. (Coleoptera), *J. Cell Sci.,* 1, 175, 1966.

509. **Smith, J. D.**, Symbiotic microorganisms of

aphids and fixation of atmospheric nitrogen, *Nature (London),* 162, 930, 1948.
510. **Soldan, T., Weyda, F., and Matha, V.,** Structural changes of mycetome in starved females of tsetse fly, *Glossina palpalis palpalis* (Diptera, Glossinidae), *Acta Entomol. Bohemoslov.*, 83, 266, 1986.
511. **Srivastava, P. N.,** Nutritional physiology, in *Aphid, Their Biology, Natural Enemies and Control,* Harrewijn, P. and Minks, A. K., Eds., Elsevier, Amsterdam, in press.
512. **Srivastava, P. N. and Auclair, J. L.,** An improved chemically defined diet for the pea aphid, *Acyrthosiphon pisum* (Harris), *Ann. Entomol. Soc. Am.*, 64, 474, 1971.
513. **Srivastava, P. N. and Auclair, J. L.,** Role of single amino acids in phagostimulation, growth, and survival of *Acyrthosiphon pisum, J. Insect Physiol.*, 21, 1865, 1975.
514. **Srivastava, P. N. and Auclair, J. L.,** Effects of antibiotics on feeding and development of the pea aphid, *Acyrthosiphon pisum* (Harris), *Can. J. Zool.*, 54, 1025, 1976.
515. **Srivastava, P. N., Auclair, J. L., and Srivastava, U.,** Effect of nonessential amino acids on phagostimulation and maintenance of the pea aphid, *Acyrthosiphon pisum, Can. J. Zool.*, 61, 2224, 1983.
516. **Srivastava, P. N., Gao, Y., Levesque, J., and Auclair, J. L.,** Differences in amino acid requirements between two biotypes of the pea aphids, *Acyrthosiphon pisum, Can. J. Zool.*, 63, 603, 1985.
517. **Srivastava, P. N., Srivastava, U., Thakur, M., and Auclair, J. L.,** Synthesis of proteins and nucleic acids by the pea aphid *Acyrthosiphon pisum* (Aphididae) in the absence of a full complement of dietary amino acids, *Arch. Insect Biochem. Physiol.*, 4, 161, 1987.
518. **Stanier, R. Y., Ingraham, J. L., Wheelis, M. L., and Painter, P. R.,** *The Microbial World,* 5th ed., Prentice-Hall, Englewood Cliffs, NJ, 1986, 113.
519. **Starmer, W. T.,** Associations and interactions among yeasts, *Drosophila* and their habitats, in *Ecological Genetics and Evolution. The Cactus-Yeast Model System,* Barker, J. S. F. and Starmer, W. T., Eds., Academic Press, New York, 1982.
520. **Steffan, A. W.,** Ectosymbiosis in aquatic insects, in *Symbiosis,* Vol. 2, Henry, S. M., Ed., Academic Press, New York, 1967, 207.
521. **Steinhaus, E. A.,** The microbiology of insects with special reference to the biologic relationships between bacteria and insects, *Bacteriol. Rev.*, 4, 17, 1940.
522. **Steinhaus, E. A.,** A study of the bacteria associated with thirty species of insects, *J. Bacteriol.*, 42, 757, 1941.
523. **Steinhaus, E. A.,** Extracellular bacteria and insects, in *Insect Microbiology,* Steinhaus, E. A., Ed., Comstock, Ithaca, NY, 1946, 9.
524. **Steinhaus, E. A.,** *Insect Microbiology,* Hafner, New York, 1967, 763.
525. **Stephens, K.,** Pheromones among the prokaryotes, *Crit. Rev. Biocompat.*, 13, 309, 1986.
526. **Stradling, D. J. and Powell, R. J.,** The cloning of more highly productive fungal strains: a factor in the speciation of fungus-growing ants, *Experientia,* 42, 962, 1986.
527. **Strong, F. E.,** Detection of lipids in the honeydew of an aphid, *Nature (London),* 205, 1242, 1965.
528. **Suic, K.,** Symbiotische Saccharomyceten der echten Cicaden, Cicadidae, *Sitzungsber. böhm. Wiss., Prag,* 3, 1, 1910.
529. **Svoboda, J. A.,** Recent developments in insect steroid metabolism, *Annu. Rev. Entomol.*, 20, 205, 1975.
530. **Svoboda, J. A., Thompson, M. J., Robbins, W. E., and Kaplanis, J. N.,** Insect steroid metabolism, *Lipids,* 13, 747, 1978.
531. **Taylor, E. C.,** Cellulase digestion in a leaf eating insect, the Mexican bean beetle, *Epilachna varivestis, Insect Biochem.*, 15, 315, 1985.
532. **Thayer, D. W.,** Facultative wood-digesting bacteria from the hind-gut of the termite *Reticulitermes hesperus, J. Gen. Microbiol.*, 95, 287, 1976.
533. **Thayer, D. W.,** Carboxymethylcellulase produced by facultative bacteria from the hind-gut of the termite *Reticulitermes hesperus, J. Gen. Microbiol.*, 106, 13, 1978.
534. **To, L., Margulis, L., and Cheung, T. W.,** Pillotinas and hollandinas: distribution and behaviour of spirochaetes symbiotic in termites, *Microbios,* 22, 103, 1978.
535. **Trager, W.,** Mitochondria or micro-organisms?, *Science,* 116, 332, 1952.
536. **Trager, W.,** *Symbiosis,* Reinhold Book Corp., New York, 1970.
537. **Tremblay, E. and Tripodi, G.,** Ultrastructural data on pseudococcid endosymbionts (Homoptera, Coccoidea), in *Endocytobiology,* Schwemmler, W. and Schenk, H. E. A., Eds., Walter de Gruyter, Berlin, 1980, 419.
538. **Tsiropoulos, G. J.,** Bacteria associated with the walnut husk fly, *Rhagoletis completa, Environ. Entomol.*, 5, 83, 1976.
539. **Tsiropoulos, G. J.,** Effect of antibiotics incorporated into defined adult diets on survival and reproduction of the walnut husk fly, *Rhagoletis completa* Cress. (Diptera, Trypetidae), *Z. Angew. Entomol.*, 91, 100, 1981.
540. **Turner, R. B.,** Dietary requirements of the cotton aphid, *Aphis gossypii:* the sulfur-containing amino acids, *J. Insect Physiol.*, 17, 2451, 1971.
541. **Turner, R. B.,** Quantitative requirements for

tyrosine, phenylalanine and tryptophan by the cotton aphid, *Aphis gossypil* (Glover), *Comp. Biochem. Physiol. A,* 56, 203, 1977.

542. **Uichanco, L. B.,** Studies on the embryogeny and postnatal development of the Aphididae, with special reference to the history of the "symbiotic organ", or "mycetom", *Philipp. J. Sci.,* 24, 143, 1924.
543. **Unterman, B. M., Baumann, P., and McLean, D. L.,** Genetic characterization and evolution of the pea aphid symbionts, 18th Int. Cong. Entomology, Vancouver, 1988.
544. **Uvarov, B.,** *Grasshoppers and Locusts. A Handbook of General Acridology,* Vol. I, Cambridge University Press, Cambridge, 1966, 79.
545. **Vago, C. and Laporte, M.,** Microscopie electronique des symbiontes globuleux des aphides, *Ann. Soc. Entomol. Fr.,* 1, 181, 1965.
546. **Veivers, P. C., O'Brien, R. W., and Slaytor, M.,** Role of bacteria in maintaining the redox potential in the hindgut of termites and preventing entry of foreign bacteria, *J. Insect Physiol.,* 28, 947, 1982.
547. **Wade, M. J. and Stevens, L.,** Microorganism mediated reproductive isolation in flour beetles (genus *Tribolium*), *Science,* 227, 527, 1985.
548. **Waterhouse, D. F., Hacknean, R. H., and McKellar, J.,** An investigation of chitinase activity in cockroach and termite extracts, *J. Insect Physiol.,* 6, 96, 1961.
549. **Weber, N. A.,** Fungus culturing by ants, in *Insect-Fungus Symbiosis. Nutrition, Mutualism, and Commensalism,* Batra, L. R., Ed., John Wiley & Sons, New York, 1979, 77.
550. **Weis, A. E.,** Use of a symbiotic fungus by the gall maker *Asteromyia carbonifera* to inhibit attack by the parasitoid *Torymus capite, Ecology,* 63, 1606, 1982.
551. **Werner, E.,** Die Ernahrung der Larve von *Potosia cuprea* Frb. (*Cetonia floricola* Hbst). Ein Beitrag zum Problem der cellulose-Verdauung bein insecten-Larven, *Z. Morphol. Öekol. Tiere,* 6, 150, 1926.
552. **West, C. A., Bruce, R. J., and Jin, D. F.,** Pectic fragments of plant cell walls as mediators of stress responses, in *Structure, Function and Biosynthesis of Plant Cell Walls,* Dugger, W. M. and Bartnicki-Garcia, S., Eds., American Soc. Plant Physiol., Rockland, 1984, 359.
553. **White, T. C. R.,** The abundance of invertebrate herbivores in relation to the availability of nitrogen in stressed food plants, *Oecologia,* 63, 90, 1984.
554. **Whitney, H. S., Mandoni, R. J., and Oberwinkler, F.,** *Entomocorticium dendroctoni,* new genus new species (Basidiomycotina), a possible nutritional symbiote of the mountain pine beetle in lodgepole pine in British Columbia (Canada).
555. **Wicker, C. and Nardon, P.,** Development responses of symbiotic and aposymbiotic weevils *Sitophilus oryzae* L. (Coleoptera, Curculionidae) to a diet supplemented with aromatic amino acids, *J. Insect Physiol.,* 28, 1021, 1982.
556. **Wicker, C. and Nardon, P.,** Differential vitamin requirements of symbiotic and aposymbiotic weevils, *Sitophilus oryzae,* in *Endocytobiology,* Schenk, E. A. and Schwemmler, W., Eds., Walter de Gruyter, Berlin, 1983, 733.
557. **Wicker, C., Guillaud, J., and Bonnot, G.,** Comparative composition of free, peptide and protein amino acids in symbiotic and aposymbiotic *Sitophilus oryzae* (Coleoptera, Curculionidae) *Insect Biochem.,* 15, 537, 1985.
558. **Wigglesworth, V. B.,** Symbiosis in blood-sucking insects, *Tijdschr. Entomol.,* 95, 63, 1952.
559. **Wigglesworth, V. B.,** The breakdown of the thoracic gland in the adult insect *Rhodnius prolixus, J. Exp. Biol.,* 32, 1955.
560. **Wigglesworth, V. B.,** Histochemical studies of uric acid in some insects. I. Storage in the fat body of *Periplaneta americana* and the action of the symbiotic bacteria, *Tissue Cell,* 19, 83, 1987.
561. **Woese, C. R.,** Archeabacteria and cellular origins: an overview, *Zentralbl. Bakteriol. Parasitenkd. Infektionskr. Hyg. Abt. 1 Orig.,* C3, 1, 1982.
562. **Woese, C. R. and Fox, G. E.,** Phylogenetic structure of the prokaryotic domain: the primary kingdoms, *Proc. Natl. Acad. Sci. U.S.A.,* 74, 5088, 1977.
563. **Woese, C. R., Stackebrandt, E., Macke, T. J., and Fox, G. E.,** A phylogenetic designation of the major eubacterial taxa, *Syst. Appl. Microbiol.,* 6, 143, 1985.
564. **Wollman, E.,** La méthode des élevage aseptiques en physiologie, *Arch. Intern. Physiol.,* 18, 194, 1921.
565. **Wood, D. L.,** The role of pheromones, kairomones, and allomones in the host selection and colonization behaviour of bark beetles, *Annu. Rev. Entomol.,* 27, 411, 1982.
566. **Wool, D., Bunting, S., and van Emden, H. F.,** Electrophoretic study of genetic variation in British *Myzus persicae* (Sulz.) (Hemiptera, Aphididae), *Biochem. Genet.,* 16, 987, 1978.
567. **Wright, J. D. and Barr, A. R.,** The ultrastructure and symbiotic relationships of *Wolbachia pipiens* of mosquitoes of the *Aedes scutellaris* group, *J. Ultrastruct. Res.,* 72, 52, 1980.
568. **Yamasaki, M.,** Studies on the intestinal protozoa of termites. II. Oxygenation experi-

ments under the influence of temperature, *Mem. Coll. Sci. Kyoto Imp. Univ. Ser. B,* 7, 179, 1931.

569. **Yamin, M.,** Axenic cultivation of the cellulolytic flagellate, *Trichomitopsis termopsidis* (Cleveland), from the termite, *Zootermopsis, J. Protozool.,* 25, 535, 1978.
570. **Yamin, M. A.,** Flagellates of orders Trichomonadida Kirby, Osymonadida Grasse and Hyperstigida Grassi & Foa reported from the lower termites (Isoptera families Mastotermidae, Rhinotermitidae, and Serritermitidae) and from the wood-feeding roach *Cryptocercus*, in *Sociobiology,* Vol. 4, Kistner, D. H., Ed., 1979.
571. **Yamin, M.,** Cellulolytic activity of an axenically-cultivated termite flagellate, *Trichomitopsis termopsidis, J. Gen. Microbiol.,* 113, 417, 1979.
572. **Yamvrias, C., Panagopoulos, C. G., and Psallidas, P. G.,** Preliminary study of the internal bacterial flora of the olive fruit fly (*Dacus oleae* Gmelin), *Ann. Inst. Phytopathol. Benaki,* 9, 201, 1970.

2 The Relative Importance of Vertebrate and Invertebrate Herbivores in Plant Population Dynamics

Michael J. Crawley
Department of Pure and Applied Biology
Imperial College
Silwood Park, Ascot, Berkshire, England

TABLE OF CONTENTS

I. INTRODUCTION

Generalizations about the relative importance of vertebrate and invertebrate herbivores in determining plant population dynamics and vegetation structure[64] have in the past been based on a mass of evidence of the profound effects that can result from exclusion of vertebrate herbivores in fencing experiments, but on only the most meager evidence concerning the effects of excluding invertebrate herbivores from natural vegetation using insecticides. There has been a substantial increase in interest on the effects of herbivory in natural plant communities in recent years, and we are now in a better position to assess the relative importance of the two herbivore groups.

A. BACKGROUND ASSUMPTIONS

1. It is clear that herbivores are more likely to have a profound impact on plant community structure if their numbers are food-limited. An enemy-regulated population would be at lower equilibrium density than one which was food-limited. It is partly because conventional wisdom suggests that herbivore numbers tend *not* to be food limited[116,201] that the belief has arisen that herbivores have little to do with the structuring of plant communities. This is a recent view, however, and it is clear from the writings of Darwin[81] and Kerner[158] that 19th century naturalists were aware of the importance of herbivory as a force in plant community dynamics.
2. Polyphagous herbivores will have a greater effect on the population dynamics of their preferred food plant species than on less-preferred species. Animals sustained by an abundant but less-preferred food will exert a very high grazing pressure on any uncommon preferred food species.[119,154]
3. Plant herbivore interactions may be highly asymmetrical. It is possible, for example, for the herbivore to be food limited but the plant not to be herbivore limited (e.g., ragwort and cinnabar moth[78,195]).
4. The recent documentation of a number of convincing cases of food limitation among both insect and mammalian herbivores[88,225] at least allows the possibility that these animals might be important in affecting the structure of vegetation.
5. In the majority of cases, however, we cannot say whether a population of herbivores is food limited, enemy limited, or even whether it is regulated at all.[66]

The potential impact of herbivory on plant abundance is determined in part by whether or not the herbivore is capable of mounting functional or numerical responses to changes in plant abundance. The responses must be sufficiently large and sufficiently rapid to check (or reverse) any change in plant abundance. Functional responses can result from switching behavior, as when the herbivore alters the composition of its diet as a result of short-term changes in relative food availability.[139] Numerical responses come about through changes in the rates of birth, death, or dispersal that result from changes in food availability. Numerical responses acting by dispersal occur when mobile herbivores aggregate in regions of temporally high food availability.[246] Longer-term responses may be due to increased reproductive success,[25,42,128,188] reduced mortality in years, or places where food is plentiful.[225,227] The question as to whether or not vertebrate and invertebrate herbivores are fundamentally different in their effects on plant dynamics can therefore be rephrased by asking whether vertebrates are more likely to exhibit strong, immediate, functional or numerical responses.

It does not follow, of course, that herbivores that are *not* food limited or that do *not* mount numerical responses to changes in plant abundance are incapable of influencing plant community dynamics. For example, a mobile, polyphagous herbivore that is regulated at low densities by natural enemies or by the availability of safe breeding places might still exclude a particular plant from the community altogether, should that species happen to be highly preferred.

A major difficulty in assessing the role of herbivory in plant dynamics lies in demonstrating the impact of herbivore feeding on plant demography rather than on plant performance. It is one thing to show that insect attack reduces seed production, but it is quite another to show that a reduction in seed production leads to reduced seedling recruitment and to a reduction in the size of the breeding population of plants in the next generation.[73,110,126,205]

In this chapter, those attributes of vertebrate and invertebrate biology that might influence their impact plant recruitment and mortality are compared, and the effects on different aspects of plant performance by the two kinds of herbivore are considered. The review is restricted to examples from terrestrial habitats, but there is a substantial literature on the impact of herbivory on marine algae,[120,122] in the rocky intertidal,[79,92,147] and in freshwater ecosystems.[22,32,167]

B. VERTEBRATE AND INVERTEBRATE HERBIVORES COMPARED

1. Body Size

Perhaps the most conspicuous difference between the vertebrate and invertebrate herbivores is their size. The largest invertebrate herbivores (e.g., a female *Locusta migratoria* weighs 1.5 g) are smaller than the smallest vertebrate herbivores (e.g., a seed-feeding pocket mouse such as *Perognathus flavus* weighs 7 g), so there is virtually no overlap in body size. Since food intake increases with body weight, the per capita rates of damage are substantially higher for vertebrate herbivores (even though larger animals take less per unit body weight, because feeding scales with body mass in an allometric fashion — with $W^{0.75}$). Clearly, a given number of vertebrate herbivores will inflict greater total damage on the vegetation than the same number of invertebrates. Body size differences also mean that vertebrates take bigger individual bites, with potentially important effects on plant performance (see below).

Large body size gives vertebrates the extra advantages of being able to extract usable energy from low quality foods more efficiently, to escape from predation, and to succeed in interspecific aggression.[87] As a result, the same amount of available energy may be able to support a greater biomass of large-bodied species.[34] Unfortunately, the range of habitats and animal groups for which good data are available to test this hypothesis is very restricted (see Section I.A.3).

2. Metabolic Rate

Homeothermic vertebrate herbivores have much higher feeding rates than poikilothermic species of similar body mass. Weight for weight, the metabolic rate of homeotherms is approximately 30 times higher.[185] Homeothermic production is only about 3% of respiration, compared with 40% for long-lived poikilotherms and as much as 83% for some short-lived poikilothermic herbivores.[135]

3. Population Density

In order to estimate the net impact of different kinds of herbivores on plant performance, their per capita feeding rates are multiplied by their population densities. In general, smaller animals have much higher population densities than larger ones. For insects, Morse et al.[193] estimate that a 10-fold reduction in body length is associated with a 178-fold increase in population density, and Peters[207] suggests that once size effects have been accounted for, terrestrial invertebrates are ten times as abundant as vertebrates. Unfortunately, there are extremely few data against which to test these predictions.

Once allowance has been made for the higher population densities of invertebrate herbivores, there is little evidence to suggest that there is any substantial difference in the impact of vertebrates and invertebrates as measured in terms of plant matter harvested per unit area. Humphreys[135] presents data on energy budgets for a variety of different animals which suggest that there is little to choose in terms of total impact. He gives respiration figures (log 10 cal/m^2/year) of 3.4 and 3.0 for Orthoptera and Hemiptera, compared with 2.9 and 3.8 for mice and voles, respectively.

Thus, the higher per capita rates of feeding of the homeothermic vertebrates, with their greater body size and higher rate of energy use per unit body weight, appear to compensate for the much greater population densities of invertebrates. In a careful comparison of abundance and body size of different vertebrate species, Brown and Maurer[34] contradict earlier studies,[80,207] arguing that the increased abundance of smaller species does not compensate for their lower individual feeding rates and that larger species utilize a disproportionately large share of the plant dry weight taken per unit area of habitat. We clearly need more high-quality data, using information collected over many years on the energetics of *all* the herbivorous species in a particular habitat, in order to resolve this question.

4. Food Specificity

The majority of herbivorous insects are monophagous or oligophagous, feeding from at most five or six plant species and often from only one. For sedentary species such as leaf miners and gall formers, the modal number of host plant species attacked per insect species in one.[7,50] For larger, more mobile insects, the modal species tends to be oligophagous, taking between two and ten plant species.[204] It is exceptional for an invertebrate species to feed on more than half of the plant species present in a community.[47] In contrast, vertebrate herbivores are typically polyphagous. They vary in the degree of polyphagy they exhibit, with larger species typically taking a wider range of foods than smaller ones.[16] Most vertebrate herbivores feed on from tens to scores of plant species (exceptions include such notoriously picky eaters as giant pandas and koala bears). The average breadth of diet therefore represents a fundamental difference between these two groups of herbivores; vertebrates are distinctly more polyphagous than invertebrates.[27,59,143,161,210]

Body-size differences also mean that feeding specificity is different at the scale of plant tissues. Thus, there are no vertebrates small enough to be leaf miners and no invertebrates large enough to fell whole trees on their own. The typical vertebrate herbivore takes many plant modules (sensu Harper[118]) per mouthful, whereas the typical invertebrate herbivore takes fractions of a module per mouthful.

It is not clear whether feeding specificity is linked with likely impact on plant performance in a predictable way. For example, we might ask whether polyphagous herbivores are more or less likely to be food limited, or whether they are more or less likely to reduce their food plant species to low densities or to eliminate them. These questions are extremely difficult, but the answer to the second is almost certainly yes. Polyphagous herbivores are more likely to reduce the abundance of at least some of their food plant species simply because their numbers are more or less independent of the abundance of scarce, highly preferred plant species.

If a monophagous herbivore were food limited, its numbers might decline before the plant population was reduced severely, or the interaction may be so strongly asymmetrical that the herbivore has no influence on plant abundance despite being food limited.[78,195] If a monophagous herbivore were not food limited, it is unlikely that it would reach population densities high enough to inflict substantial losses on its host plants. Most of the examples where monophagous herbivores *have* reduced host plants to stable, low densities come from biological weed control. In such cases, both the plant and the herbivore are introduced to foreign plant communities where the herbivore is freed from its natural enemies by careful, prerelease screening. Even in these cases, however, it is not clear that the insect responsible for the initial reduction in weed density is currently responsible for the maintenance of stable, low-density weed populations because so many other things are changed following successful weed control (e.g., land management practice is often improved). The acid test would involve spraying insecticide on the areas of successful weed control. If the weeds really were regulated at low density by the introduced insects, then the weeds should return to their former abundance on the sprayed plots. Needless to say, this experiment has never been attempted.[73,163]

5. Mobility

There is little difference between vertebrate and invertebrate herbivores in terms of their lifetime mobility. Tiny insects such as aphids are capable of flight over many thousands of kilometers[236] and large ungulates such as wildebeest migrate over similarly great distances.[179] At the other extreme, sessile insects may spend many successive generations on a single individual plant.[94,244] Some vertebrate herbivores are also rather sedentary, with many successive generations occupying the same home range.[256] Even the smallest territories, however, contain hundreds or thousands of individual plants.

The immature stages of most invertebrate herbivores are restricted in their mobility and, while they may be able to walk from one plant to another, long distance dispersal is frequently impossible. In contrast, juvenile vertebrate herbivores can move alone or with their parents over considerable distances. Thus, while there is little to choose between the mobilities of the adult stages, vertebrate herbivores are mobile for a greater proportion of their life span. The result is that both habitat selection and food plant selection within habitats can occur continuously throughout the lifetime of an individual vertebrate herbivore. This means that discovery of resource-rich patches and aggregation within these patches (i.e., numerical responses through dispersal) are more likely to be exhibited by vertebrates, particularly when the food patches are small and easily depleted. Mobile flocks of polyphagous seed-feeding birds, for example, may have a more profound effect on seedling recruitment than resident, monophagous invertebrate seed feeders[181,246] whose numerical responses occur only through reproduction and entail a time lag of at least one generation.

6. Starvation Tolerance

Vertebrates and invertebrates differ in the absolute size of their bodily food reserves and hence in their responses to food depletion. Whereas an invertebrate may be capable of entering facultative diapause or obligate diapause rapidly in response to starvation, a homeothermic vertebrate may continue to deplete the plant population long after its net energy gain from grazing has become negative. This kind of starvation tolerance can cause very high rates of plant exploitation by vertebrates, and theoretical models embodying this feeding behavior are highly unstable.[63]

Many vertebrates are more buffered against declines in food quality than are their invertebrate counterparts. Their larger body size, more complex gut floras, and higher power of mastication suggest a greater tolerance of reduced food quality. This, in turn, means that vertebrates are likely to have a greater impact on plant performance because their feeding declines more slowly (or after a more pronounced time lag) following the onset of a decline in food quality.[13] Insect herbivores generally are extremely sensitive to reductions in food plant quality.[4]

Response to starvation affects the way that negative numerical responses (population declines) occur in the two groups. Emigration is the immediate response to declining food availability by almost all herbivores. Once there is nowhere left to go, however, herbivores that have no dormancy mechanisms (such as the ungulates with their large body size and rather limited ability to slow metabolic functions) tend to go on depleting plant resources until they starve to death.[171] Facultative diapause may allow invertebrate herbivores to survive temporary, local depletion of food availability without inflicting such severe depletion on their host plants.[226]

7. Increasers and Decreasers

In the language of range management, plant species that become more abundant in heavily grazed pastures are known as increasers and plants preferred by the herbivores, or intolerant of defoliation, that decline in abundance are known as decreasers. If a fence is erected to exclude vertebrate herbivores, the process is reversed; the increasers decline in abundance because of competition from the taller, usually more vigorous decreasers (see below).

8. Ice Cream Plants

Some plants are so attractive that herbivores eat them whenever they are found. These "ice cream plants" represent an extreme form of decreaser and are so attractive that they may be excluded altogether from the plant community. Vertebrate herbivores tend to exhibit this behavior more often than invertebrates because of their greater mobility and higher polyphagy. Numerous experiments with sheep, rodents, and rabbits have demonstrated this process in operation.[3,71,84,213] However, there has been little direct experimental work to determine the rate at which "ice cream plants" are eliminated from a plant community following the introduction of herbivores. A 5-year study of rabbit grazing on ruderal plant regeneration following cultivation of acid grassland suggests that (1) most of the species in the seed bank beneath undisturbed grassland are decreasers that are absent from the vegetative community and (2) both the number of ruderal species and the proportion that are increasers increase with the frequency of disturbance.[71] An experiment carried out in a large, heterogeneous sheep paddock in Australia involved placing isolated cubes of commercial sheep feed at very low densities in a range of microhabitats. The cubes all disappeared rapidly, even from areas where grazing pressure was thought to be negligible.[252a]

Mobile polyphagous invertebrates are also capable of eliminating "ice cream plants", as Parker and Root[206] describe regarding the exclusion of *Macaeranthera* by grass hoppers. The insects are maintained at high density by feeding on *Gutierrezia,* the dominant shrub within the community. Even oligophagous lepidopteran caterpillars can inflict high rates of mortality. For example, the checkerspot butterfly, *Euphydryas editha,* kills *Collinsia* plants in the vicinity of its preferred host plant *Pedicularis semibarbata.*[238] Spruce budworm, *Choristoneura fumiferana,* can eliminate balsam fir, *Abies balsamea,* from mixed forests in outbreak areas.[192] Other invertebrate herbivores such as slugs have been shown to exhibit frequency-dependent switching in their feeding behavior, preferentially attacking the most abundant host plant. This provides a refuge, reducing the probability of extinction of the rarer plant.[60]

9. Predictions

Taken together, these differences lead to the prediction that vertebrate herbivores would exert a greater influence on plant community structure and dynamics. In particular, their greater body size, individual bite size, mobility, starvation tolerance, and polyphagy suggest that exclusion of vertebrates would have both a more immediate and, in the long term, more profound impact on plant populations.

C. PLANT PERFORMANCE INDICATORS

Feeding by herbivorous animals almost invariably reduces plant performance and presumably plant fitness.[18,69] In the very small number of cases where feeding has been shown to increase plant performance,[29,138,202] it is not clear what impact, if any, herbivore attack has on plant abundance in the next generation. In no case has herbivore feeding been shown to enhance the genetic fitness of the grazed individuals.

The indicators of plant performance considered here are (1) flower production, (2) fruit production, (3) seed production, (4) seed size, (5) seed predation, (6) seedling survival, (7) rate of growth and vegetative spread, and (8) death rate of mature plants. Detailed reviews of the impact of herbivory on these different aspects of performance are to be found elsewhere.[64,74] The purpose of this section is simply to highlight a number of differences between the effects of vertebrate and invertebrate herbivory.

1. Flower Production

Both vertebrate and invertebrate herbivores affect flowering directly by eating floral meristems, as well as indirectly by defoliation, sap sucking, stem boring, or root feeding, all of which reduce the availability of resources necessary for flower production. For example, rabbit grazing markedly depresses flowering in *Hieracium pilosella,*[26] in wheat,[75,76] and in many pasture grasses (notably *Festuca rubra*).[71] Vertebrates are important predators of flower buds. Small mammals may feed preferentially on nutrient-rich flower buds,[98] and some of the worst pests of flowers include birds such as finches.[234] Indirect effects of herbivory also act through reduced plant size; individuals that are kept small by defoliation almost always flower substantially later, and the smallest plants may never flower at all before they die.[162]

Invertebrates can also have a marked impact on flowering. Heavy attack by the grasshopper *Hesperotettix* casues an 80% reduction in the number of bushes of *Gutierrezia* that flower,[205] while the grass aphid *Holcaphis holci* greatly reduces the proportion of tillers of *Holcus mollis* that produce flowering shoots.[201a] A number of invertebrates destroy flowers, steal nectar, and otherwise interfere with pollination.[115,158]

2. Fruit Production

Unripe fruits tend to be well defended against herbivores and contain high concentrations of toxins or feeding deterrents.[142] Even so, rates of herbivory at this stage can be very high.[28] In coniferous woodland, for example, squirrels may take up to 80% of cones before ripening,[20] and spruce budworm can account for a 40% cone loss during the period May to August, inflicting less serious damage (approximately 22%) on the surviving cones.[89]

It is often difficult to attribute losses of unripened fruit unequivocally to herbivore feeding, since natural abortion accounts for a substantial loss of fruit, even in plants that are not attacked.[33,231] Herbivory may therefore act to exaggerate what are already rather high losses or, conversely, may account for fruits that would have been aborted in any case.[163] Thus, there is substantial scope for compensation when fruit loss to herbivores leads to a reduction in the rate of natural fruit abortion.[230]

3. Seed Production

Total seed production is likely to be affected by the timing of attack and by the overlap between the feeding period of the herbivore and the period over which the plant accumulates the resources necessary for fruit production and seed fill. Thus, plants subject to attack by short-lived, univoltine insects[139] tend to suffer less than plants attacked throughout their growing period by multivoltine insects.[220] In seasonal environments, most plants are exposed to vertebrate herbivory for the entire course of their development, but to invertebrates for only limited, often rather brief periods. Even in supposedly aseasonal tropical environments, there are numerous univoltine insects.[143]

Because most plants are exposed to vertebrate herbivory for a greater portion of the growing period, and vertebrate herbivores are somewhat less selective in the ages of leaves they will eat, it would be expected that they have a greater depressive effect on seed production. In the few cases where there are direct, comparative data, this appears to be the case. Wheat loses significantly more seed production as a result of defoliation by rabbits as compared to attack by cereal aphids and other insects.[72] Flowering of the grasses *F. rubra* and *Agrostis capillaris* is much higher on fenced plots than on plots grazed by rabbits, but no higher on insecticide-sprayed plots.[71] Insects such as aphids and stem-boring flies can, however, greatly reduce grass seed production in some habitats.[37,52,111] A synthesis of the data relating herbivore feeding to seed production has never been attempted, but it would almost certainly show different patterns in the relative importance of vertebrate and invertebrate herbivory in different plant communities (e.g., insects are probably the most important determinants of seed production by certain forest trees).[65,146]

4. Seed Size

The notion that seed size is a species-specific constant has been dispelled by the mass of evidence gained by weighing individual seeds.[70,124,139,220,239] If bulk seed is weighed (e.g., the practice of determining 1000-seed weight), then obviously the variance in estimated seed weight decreases. (It is like comparing the standard deviation of a sample — s — with the standard error of the mean — $s/n^{1/2}$.) Substantial variation in seed weight has been shown to result from defoliation by vertebrate[76] and invertebrate herbivores.[157,162,190,239,257] These smaller seeds often produce smaller, less-competitive seedlings that are more likely to die before flowering,[49,77,105,196,219] but Hendrix[123a] has not found reduced performance with smaller seeds of *Pastinaca sativa.*

Removal of ripening fruits by herbivores, on the other hand, may mean that the surviving seeds are larger.[46] Notice, however, that the standard horticultural practice of removing young fruits so that the surviving *fruits* are larger does not necessarily lead to the production of larger individual seeds. The precise position of a seed within a fruit can be important in affecting both its size and its dormancy characteristics;[97,257] both can be influenced by herbivory.

The survival of the embryo, despite the removal of part of the endosperm, may be possible when small invertebrate seed feeders attack relatively large seeds. Experimenting with the acorns of *Quercus robur,* Forrester[104a] has found that surgical removal of up to 70% of the cotyledon mass still allows 20% seedling establishment. Individual oak trees are consistent in the size of acorns they produce, with the largest acorns weighing 7 times as much as the smallest. Thus, feeding damage by curculionid weevils may cause lower death rates on trees that produce larger seeds, even though the percentage of fruits attacked is the same for trees with different sizes of fruit.

5. Seed Predation

As a rule, vertebrate predators of seeds tend to be polyphagous, feeding on many species of seed and taking other plant parts at different times of year, whereas invertebrates are often restricted to feeding only on seeds taken from one or a few plant species.[1,9,83,143] This means that vertebrate seed feeders are more likely to be able to mount rapid functional and dispersive responses to local changes in seed abundance, while invertebrates are more likely to exhibit time-lagged numerical responses.

No exhaustive compilation of data on rates of postdispersal seed predation has ever been attempted. Superficial analysis might suggest that vertebrate herbivores cause higher average rates of seed predation (of the order of 40%[85,186]) than invertebrates (roughly 20% seed mortality[2,240]), but the variation from study to study is so great that generalization is impossible. Experiments with the acorns of *Q. robur* showed 100% seed mortality in each of 6 years, irrespective of the size of the cache of seeds or of the distance the seeds were placed from parent oak trees.[71] Buried acorns suffered only 20% mortality in the same experiments,[104b] demonstrating the importance of seed burial by one set of seed predators (mice and jays) in protecting seeds from potentially higher rates of predation by other herbivores.

Insect species can also inflict extremely high rates of seed mortality. For example, the cynipid gall wasp *Andricus quercuscalicis* can kill 100% of the acorns on certain trees of *Q. robur* and, in most years, kills a larger number of acorns than are eaten by mice and rabbits.[68] Carabids may be more important predators of umbellifer seed than mice,[164,240] and bruchid beetles kill a great many more seeds than are taken by deer in tropical forests (95 vs. 20%[144]).

Some kinds of postdispersal seed feeding are hard to detect and therefore extremely difficult to quantify (e.g., predation by seed-feeding lygaeid bugs that make only a tiny hole in the seed coat[57,243]). Such feeding leads to underestimation of the seed mortality attributable to invertebrate herbivores; losses of this kind would typically be scored as "inviable seed".

The relationship between mast fruiting and seed predation rates has received considerable attention in recent years.[141,223,245] The hypothesis is that large seed crops are produced simultaneously over extensive geographic areas, followed by more or less protracted periods in which there is virtually no seed production. This pattern of fruiting means that specialist seed predators occur at low densities and are readily satiated in the mast year. The process is assumed to work because neither specialist nor generalist seed predators are able to mount sufficiently rapid numerical responses in the period between seed fall and germination to inflict high percentage losses (i.e., predation rates in masting species are assumed to be negatively density dependent). The data, however, are equivocal. Clearly, predator satiation is a potentially important determinant of seedling recruitment, but in a number of cases where density-dependent seed survivorship has been sought, it has not been found. Seeds were just as likely to be eaten when they were part of a large seed crop as when part of a small crop.[132] In a few cases, seeds suffered a *higher* death rate in large crops because of herbivore aggregation.[229,232] Results consistent with the predator satiation hypothesis have been observed with ants[2,198] and some small mammals.[149] The average efficiency of seed dispersal has been found to be lower in a mast crop,[12,82] and this might provide a potential counterselection to the evolution of masting behavior. A major practical difficulty is that it is impossible to do realistic field experiments on masting. In order to satiate seed predators over a meaningful area, the density of seed would need to be increased experimentally on plots of perhaps hundreds of hectares. Simply obtaining enough seed to do this (especially in a nonmast year) would present a formidable problem, not to mention either the costs or the difficulties of monitoring seed loss rates over so large an area. It is clear, however, that mast fruiting is a more effective defense against invertebrate herbivores than against vertebrates, with their higher polyphagy and ability to aggregate rapidly in areas of high seed abundance.

Many seed predators also serve as the main agents of seed dispersal for the plant.[20,21,57,118,126,145,164] Thus, for species such as squirrels, jays, and harvester ants, the costs and benefits of seed consumption are exceptionally difficult to compute. The occasional acorn that is transported many hundreds of meters, yet survives to give rise to a plant that produces seed for several hundreds of years, is presumably worth a great many acorns that might be killed beneath the parent tree.[48,112] Herbivores that cache seeds but fail to rediscover all of them can be important agents of seed dispersal.[150,228] Invertebrate herbivores may affect the probability of seed dispersal by vertebrates, as when hawthorn fruits damaged by the larvae of tephritid flies are less attractive to blackbirds[180] or when weevil-infested acorns are rejected by seed-caching woodmice.[78a]

Different faunal regions differ in the relative importance of vertebrate and invertebrate seed predators. For example, in the arid southwestern U.S., the major granivores are rodents, with ants in second place and birds relatively unimportant.[35,218] In similar climates in Australia, however, there are no rodents and ants act as the major seed dispersers. This means that seed stocks are depleted more slowly and seeds survive longer in the soil in Australia.[194] The benefits that were traditionally thought to accrue from ant dispersal (e.g., transport to a nutrient-enriched microsite) have been called into question by recent experimental studies.[21,129,215]

6. Seedling Mortality

Vertebrate herbivores are capable of excluding plant species from communities by selective predation of their seedlings (e.g., deer,[165] pocket gophers,[114] and rabbits[71,169,182]). There are numerous avian pests of seedling crops, including wood pigeons, jackdaws, and skylarks.[117] Even plants such as grasses can suffer severe seedling predation by vertebrates.[211] Bark removal from young saplings by rodents may prevent natural regeneration of certain tree species, even though the adult trees produce abundant seed.[208]

In many plant communities, the major invertebrate seedling predators are slugs and snails,[44,90,91] even in deserts.[258] Mollusks can also be devastating pests to the seedlings of certain agricultural crops.[153] Insect seedling predators include sawflies,[153] weevils,[111] and mirid[109] and pentatomid bugs.[241] The major effect of insect feeding on seedlings, however, is to reduce their vigor and competitive ability,[189] rendering them more likely to die from desiccation, shading, root competition, or fungal disease. In very few cases, however, has experimental exclusion of seedling-feeding insects been demonstrated to increase the population density of flowering plants in natural vegetation.[36,37,111,175] Whether this means that their impact is negligible (because of compensatory reductions in seedling mortality through other causes) remains to be seen.

7. Rate of Growth and Vegetative Spread

Defoliation and sap feeding almost always reduce plant growth rate, with the degree of growth loss determined by the duration, timing, and intensity of herbivore feeding. Most studies report an approximately linear decline in growth with herbivore feeding,[31,75,191,203] and a few record an accelerating (quadratic) damage function[190] suggestive of some degree of compensation at low levels of attack.[64,184] Very few cases of overcompensation (where plant performance is actually higher following herbivore attack) have been convincingly documented.[17,18,69]

Plant tissues do not suffer equally following defoliation, and there is a predictable hierarchy in which the performance of different tissues (roots, leaves, stems, and fruits) is lost. It appears to be quite general, for example, that defoliation leads to reduced (or stopped) root growth, and that root pruning leads to reduced shoot production or premature abscission of above-ground parts.[217] It is less clear whether vegetative growth or seed production suffers first under herbivore attack and the limited evidence is contradictory. For example, shoot-feeding insects on oaks in Britain cause substantially reduced seed production without having any measurable effect on wood production.[65] In contrast, cicadas feeding on root xylem of oaks in the U.S. caused a 30% reduction in wood growth but had no effect on acorn production.[156]

Herbivores also affect growth indirectly by reducing the competitive ability of the individual plant relative to its neighbors. For example, grazing caused a 95% death rate of grass tillers at high plant density, but only 20% when plant competition was slight.[6] In some cases, a plant may benefit from the presence of its competitors if these other plants provide physical protection from otherwise damaging levels of herbivory.[121,182]

Invertebrate herbivores, as a result of their small size, often damage only small parts of an individual leaf in any bout of feeding. Nevertheless, their effects on leaf area are often exacerbated because feeding leads to premature leaf fall.[155] This effect is observed with some leaf miners,[39,100,130] gall formers,[254] and aphids.[260] If there is substantial premature leaf fall, it is impossible to estimate annual rates of leaf damage from a single leaf sample taken late in the season.[108,200] Again, if leaf turnover rates are high, accurate estimation of leaf consumption rates requires numerous samples taken at sufficiently brief intervals throughout the growing season to allow damage to be measured before the leaf falls.

The physiological condition of the plant can affect both the rate of herbivory to which it is subject and the consequences of a given level of herbivore feeding on plant performance. Invertebrate herbivores, for example, grow more quickly and produce more fecund adults on plants that have been stressed in a variety of ways. Thus, plants damaged by air pollutants,[103] drought,[23,178] or mechanical breakage[93,127] were preferred by the insects, and insect colonies on these plants grew more rapidly and achieved greater population densities. This response is not universal, and there are examples of insects doing better on unstressed than on stressed host plants.[248] Furthermore, it is not clear on theoretical grounds that stressed plants should be less well defended against herbivores. Simple models of optimal defense theory suggest just the opposite; there are plausible circumstances in which plants would do best by diverting *more* resources to defense when stressed.[102]

Given the size of the literature on plant stress and insect herbivores, it is surprising that so little work has been published on the effects of drought and pollution on the performance of vertebrate herbivores. No references can be found regarding physical or chemical stresses affecting the probability that a plant will be attacked by vertebrate herbivores, nor any suggestion of a more severe impact for a given level of feeding on stressed plants. It is known that vertebrate herbivory can itself bring about substantial changes in subsequent food quality. Regrowth shoots produced by certain plants following browsing or defoliation are higher in food quality,[62] while in others they are lower.[38]

In addition to the obvious effects of herbivory in reducing plant growth rate, it should be noted that plant growth rate can also affect the rate of herbivory. Slow-growing plants may be more likely to be discovered and subsequently attacked by herbivores. Within species, slow-growing plants may be less vigorous and consequently more susceptible to herbivore attack (e.g., large, slow-growing trees are most susceptible to bark beetle attack;[233] trees in dense stands are less resistant to herbivore attack[247]) or may suffer more when they are attacked (e.g., slow growing monocarpic *Lobelia* are killed by hyrax browsing, while faster growing individuals survive[259]). Between species, plants with different life histories and from environments with different rates of resource supply tend to differ in their average leaf longevities (plants adapted to resource-poor environments have long-lived leaves[55]). Long-lived leaves tend to be better equipped with antiherbivore defenses,[214] so the average *rate* of defoliation tends to decrease as average leaf longevity increases. Whether or not the total, lifetime leaf-area loss differs with leaf longevity is an important question for which there are too little data to generalize at present.

These within- and between-species differences in herbivore susceptibility can give rise to some curious and counter-intuitive correlations. For example, comparing individual oak trees may lead to the conclusion that higher growth rates are associated with higher rates of defoliation.[242] In a comparative study of light-gap and shade-tolerant tropical tree species, there was a similar positive correlation between growth rate and defoliation rate.[54] Such relationships may be expected wherever the costs of defoliation are less than the costs of increased antiherbivore defense.

8. Death Rate of Mature Plants

It is unusual for herbivores of any kind to inflict obvious mass mortality on their host plants, but there are conspicuous exceptions. With vertebrate herbivores, destruction of woodland by elephant[255] and ringbarking of trees by small mammals[208] have been observed. With invertebrates, mass mortality of trees occurs during outbreaks of bark beetles,[24] following repeated defoliation by gypsy moths,[168] and with insects that are vectors of plant diseases.[43] In many of these cases, however, it could be argued that herbivory was just the final straw and that it killed trees that were close to the end of their natural lives.

In most habitats, most of the time, the majority of plant deaths are simply not noticed because they happen to seeds and seedlings. As already discussed, the relative importance of vertebrate and invertebrate herbivory at these stages is not known. The risk of death is typically size-specific, with small plants much more likely to die than large ones. Thus defoliation, by reducing plant growth rate, keeps plants in small size classes for a longer period and exposes them to a greater net risk of death.[123,130,166,259] In some cases, herbivores themselves are size-specific in their attack rates. Stem borers, for example, may require a minimum basal shoot diameter before they can survive inside a shoot.[220] Animals may preferentially attack the largest individual plants, as was the case in Parker's study of the cerambycid beetles attacking shrubs of *Gutierrezia.*[205] In this same system, grasshoppers killed 47% of individual shrubs in dense vegetation, compared with only 22% of those whose neighbors had been removed.

II. OBSERVATIONAL EVIDENCE

A. HERBIVORE OUTBREAKS

The argument that "the world is green" has been used to support the suggestion of Hairston et al. that herbivores are enemy-limited and consequently have rather little impact on plant performance.[116] There are at least two flaws in this argument. The first lies in the assumption that scarce herbivores can have no impact on plant population dynamics. Only by excluding herbivores and comparing plant population dynamics under herbivore-free conditions can this claim be substantiated. The second lies in the implicit assumption that herbivore populations cannot be regulated by low food quality in the midst of an apparent sea of plenty. In fact, there are numerous examples of food-quality limitation.[64]

Nevertheless, there is some evidence in favor of the hypothesis of Hairston et al. It is true that herbivore outbreaks are scarce in most habitats, and the devastation caused to food plants when outbreaks *do* occur strongly suggests that the herbivores are not limited by the availability of food most of the time. Again, there are numerous examples in which removal of natural enemies has led to dramatic increases in herbivore population density.

There is a wealth of anecdotal evidence concerning the resurgence of pests following application of broad-spectrum insecticides. New pest species have been created in cases where the natural enemies suffered a more severe setback from pesticide application than did their herbivorous prey (perhaps because the natural enemies were more mobile and had higher pick-up rates of insecticide[5]). Exclusion studies, using nets to keep insectivorous birds out of trees, have shown considerable increases in caterpillar population densities and average levels of defoliation.[101]

Examples of vertebrate outbreaks caused by predator removal are far fewer (and, if anything, even more anecdotal).[56] It has been claimed that some ungulate populations are regulated by mammalian carnivores (e.g., Messier and Crete[187] suggest that moose are regulated by wolf predation), and it is well known that diseases like rinderpest and anthrax are capable of keeping ungulate populations below the level at which they would be limited by food.[224] An important role for natural enemies has recently been suggested in the regulation of several vertebrate herbivores that were previously regarded as clear-cut examples of food-limited populations (e.g., gut nematodes in red grouse population dynamics[133,209]). In cyclic populations of small mammals in the Arctic, predators are almost certainly important in determining the duration, if not the initiation, of the cycles.[125,128,148,152,173]

The evidence from population outbreaks is not strong, but it does suggest that enemy regulation is more common among invertebrate herbivores. This, in turn, adds weight to the hypothesis that vertebrate herbivores have a greater impact on plant population dynamics and vegetation structure, because the growth of their populations is more often limited by food availability.

B. INTRODUCED VERTEBRATE HERBIVORES

When vertebrate herbivores have been introduced by man onto islands (e.g., goats, rats, and reindeer) or continents (e.g., rabbits and horses), their impact on native vegetation has been profound.[30,45,99] The broad diet of the vertebrate herbivore, coupled with a lack of any evolved defenses against large vertebrate herbivores among native island floras, made these ecosystems particularly vulnerable. While it is true that introduced insects have occasionally devastated native vegetation (e.g., the alien termite *Coptotermes* threatens the endemic tree *Acacia koa* and the twig borer *Xylosandrus* threatens several other rare trees on Hawaii[131]), the restricted host-plant range of most insect herbivores has meant that the majority of accidentally introduced herbivorous insects perished at once. A few may persist at low densities with very little impact on their adopted host plants.

The comparison of the relative impact of alien vertebrate and invertebrate herbivores on native plants leaves no doubt as to the greater impact of vertebrates, but provides rather weak evidence for any generalizations about the relative impacts of the two groups of herbivores in their native environments. This is chiefly because the introduced vertebrates start out with two built-in advantages over introduced invertebrates: (1) they are more polyphagous, and hence more likely to be able to survive on the vegetation they happen to find in an alien habitat, and (2) because the plant and herbivore share no evolutionary history, the specialized chemical cues that monophagous insects often require before they will begin to feed (the so-called phagostimulants) will usually be lacking from the plants of alien floras.

C. OVERGRAZING

Overgrazing by vertebrates is recognized, but not overgrazing by invertebrates. Mismanagement of stocking rates for domestic vertebrate herbivore populations has led to severe downgrading of pasture quality in many parts of the world, especially in arid and semiarid regions.[197] Devastation of natural plant communities by insects, however, is perceived as much less common and is associated only with rare catastrophes such as locust plagues[212] or outbreaks of forest insects (these may be alien species such as gypsy moth[168] or natives such as spruce budworm[89]). This asymmetry is hardly surprising, however, since overgrazing is essentially a management problem. Overgrazing by wild vertebrate herbivores only occurs following habitat destruction (usually by humans), leading to range contraction by the animals.[17,221] Also, insect herbivores are not managed as domestic livestock. If they were, there is no doubt that insect overgrazing would be a good deal more prevalent.

In natural plant communities, chronic high rates of defoliation are sometimes associated with vertebrate grazing. This leads to the creation and maintenance of characteristic, low-stature plant communities known as "grazing lawns" that are produced by species such as the giant tortoises on Aldabra[53] or by large ungulates in the Serengeti.[17] To call these natural grazing lawns overgrazed, however, is both misleading and unnecessary.[64,184,252]

D. SUCCESSION

Most models of succession view herbivory as an external process rather than as an integral component of ecosystem dynamics.[95] Thus, vertebrate herbivores are typically seen as "delaying" or "deflecting" successions.[235] The role of invertebrate herbivores in altering the rate and direction of successional change has not been investigated at all until recently.[36,183] It is clear from numerous range management studies[96] and from exclusion of species such as sheep and rabbits from grasslands[134,213] that vertebrate herbivores can prevent the establishment of tree cover, and are therefore capable of slowing (or stopping) secondary succession. Studies with other herbivore species largely support the view that herbivory retards succession.[10,151,251] There are exceptions, as when herbivores weaken a dominant plant species that had previously been inhibiting successional change.[183] There are no reports of herbivores delaying succession as a result of their feeding preferentially on a plant species that would otherwise play a facilitating role (e.g., a nitrogen-fixing legume). Nevertheless, this process is of considerable theoretical interest and may have significance in some primary successions on very young soils. Experimental studies of succession (see below) all point toward a greater role for vertebrate than for invertebrate herbivores, reinforcing the mass of observatonal evidence.

E. THE DISTANCE-DENSITY HYPOTHESIS

A focus of research into the impact of herbivory on plant community structure has been the so-called "distance-density hypothesis" proposed independently by Janzen[140] and Connell.[58] They argued that plant recruitment should peak some distance away from a parent plant, because herbivore attack on seeds and seedlings is expected to decline with distance from the parent and with the density of seeds on the ground. Separating the effects of distance and density and determining which is the most important has proved to be something of a challenge.[8,15,143,144] It is difficult experimentally to increase seed densities to high levels over realistically large areas. Most studies, therefore, have concentrated on distance effects. The results are equivocal; some studies have shown that plant loss rates do decrease with distance from the parent plant, as expected,[8,199] but others have shown no apparent distance effects.[51] Invertebrate herbivores do appear to be more likely to exhibit distance-density effects, however, than the more mobile, more polyphagous vertebrate herbivores.

III. EXPERIMENTAL EVIDENCE

The acid test of the relative importance of vertebrate and invertebrate herbivores in determining vegetation dynamics is to carry out manipulative experiments in which the population densities of vertebrate and invertebrate herbivores are altered singly and in combination. Unfortunately, these experiments have never been carried out in natural vegetation. Even agricultural examples are scarce.

In the few cases where the density of one group of herbivores has been varied and the consequences of the manipulation on the abundance of the other group has been monitored, the clear consensus is that vertebrate herbivores have more effect on the invertebrates than vice versa.[137,160] These interactions are typically mediated via the host plant (through changes in abundance, quality, or phenology) rather than through any direct, behavioral interference between the animals (although Fowler and Saes[106] have presented an intriguing example of interference competition between ants and cattle).

In natural vegetation, most experiments have been confined to observing the consequences of removing just one of the herbivore groups by some kind of exclosure technique. Fencing against vertebrate herbivores has a long history and a rich literature.[11,14,113,136,174] The use of insecticides to exclude invertebrate herbivores from natural plant communities is a much more recent technique, and published examples are still scarce (but see below). Several long-term studies are currently in progress, however.

A. FENCING EXCLOSURES

Erecting a fence to keep out vertebrate herbivores almost always has a marked effect on the vegetation, even within the first growing season. Average plant size increases, plants which previously had not bloomed come into flower,[237,249] and species produce seedlings that had never recruited from seed in the grazed sward.[26,41,113,174,182,216,250] In the longer term, plant species richness may decline as a consequence of larger individual plant size, coupled with the competitive exclusion of some low-growing, grazing-tolerant species (e.g., prostrate and rosette plants).

In the language of range-management, fencing usually means that increasers are replaced by decreasers (i.e., species tolerant of herbivory or ignored by herbivores that thrive in heavily grazed pastures are replaced by species that are intolerant of defoliation, but are competitive — usually taller — in ungrazed conditions). The classic model of grazing dynamics envisages grazing and succession to be opposed in their effects on the direction of vegetation change. Thus, the vegetation adopts a more juvenile aspect if grazing is increased, while succession moves to a more mature phase if grazing is relaxed. In this simple model, the dynamics are fully reversible and prudent grazing management can maintain the plant community at any desired successional stage.

The classical model of succession fails to describe a number of important cases. In semiarid rangelands, for example, the increaser species are often long-lived, woody plants that persist, sometimes as virtual monoculture, for many years after fencing.[253] Similarly, the long-lived grass *Nardus stricta,* which increases on poorly managed hill grazings in Britain, persists as the dominant plant for many years after fencing has excluded the sheep.[213] The vegetation changes brought about by vertebrate grazing in these systems appear to be irreversible, at least in the medium term.

It is unfortunate that, until recently, the dynamics of vegetation change following fencing have not been studied in detail. Almost nothing is known about the rates of birth, death, and dispersal of plants inside and outside the fence. The consequences of fencing are known, but the processes by which the changes came about (e.g., the relative importance of competitor release, grazing tolerance, and herbivore preference) are not understood.

B. INSECTICIDE APPLICATION

It is only in recent years that experiments have been carried out on the exclusion of insects from natural vegetation by means of chemical pesticides. The vast literature on insecticide treatment of crop plants tells us virtually nothing about the role of insect herbivores in plant population dynamics. Something is learned about how insects affect the performance of individual crop plants, but nothing about how these changes in performance are translated into changes in plant recruitment. We need to understand the mortality factors that act at each stage of the life cycle of the plant and how these are affected by invertebrate herbivory. We need to know which factors are responsible for changes in abundance from year to year (the key factors) and which, if any, are responsible for population regulation (the density-dependent factors).

The first insecticide experiments in natural vegetation were either small scale or never published in detail. Others were disappointing, in that the prime response to insecticide treatment came about as a result of direct phytotoxic effects of the pesticide on plant species. This allowed competitor release of prevously suppressed species that were not susceptible to the phytotoxic effects of the insecticide.[222] These results gave the superficial impression that these increasing species had previously been kept scarce by insect herbivory. Subsequent experimenters have been more careful in their choice of insecticides.

Some long-term insect exclusion experiments have been carried out in seminatural vegetation. For example, exclusion of insects (mainly frit flies, *Oscinella* spp.) from grasslands has shown increased persistence of sown pasture grasses and reduced weediness. This suggests that insect herbivores can shift the competitive balance away from grasses towards herbs, thereby reducing dominance and enhancing plant species richness.[52]

By spraying shrubs of *Haplopappus squarrosus* and *H. venetus* with insecticide, Louda[175,176] found that although seed production was increased in both species, plant recruitment was increased only in *H. squarrosus.* Recruitment in *H. venetus* did not appear to be seed-limited, so insect feeding did not influence plant population density.

Sheppard[220] excluded ray-feeding aphids from flower heads of the umbellifer *Heracleum sphondylium* using insecticides and observed substantially increased seed production as a result. In a separate experiment at the same site in the same year, he sowed extra hogweed seeds and demonstrated that plant recruitment was seed-limited. These two results suggest that aphid feeding was depressing plant recruitment. In a similar experiment, Duggan[91] found no increase in recruitment of the crucifer *Cardamine pratensis* as a result of insect exclusion because of a compensatory increase in pod predation by slugs on the insecticide-sprayed plots. Where both slugs and insects were excluded, plant recruitment was enhanced, but only on plots from which the perennial plant cover had been removed experimentally. Excluding herbivores had no effect on plant recruitment in the intact wet meadow vegetation.

Insect exclusion from weedy crops of winter wheat had no measurable impact on the biomass or species composition of the weed flora. Although there was a slight increase in wheat yield in response to the removal of cereal aphids, this yield response was only measured on plots that had also been treated with an extra herbicide application.[72] No differences in the abundance of any plant species were observed after 1 year in a comparison between sprayed and unsprayed split plots that were part of a factorial experiment with and without rabbit grazing and with and without cultivation of acid grassland.[71] There may, of course, be effects that show up only after long-term insect exclusion.

For example, spraying heather, *Calluna vulgaris,* for 3 years at two sites in southern England reduced the abundance of above-ground herbivorous insects dramatically, but led to no increase in biomass, flowering, or seed production until the third year. It is noteworthy that the response to insect exclusion was only half as great as the response to a single dose of NPK fertilizer.[33a] This kind of lagged response shows that long-term study is essential, especially with slow-growing plants like *Calluna.*

There are some data to suggest that insect herbivores may be responsible for competitor release of previously suppressed plant species. For example, insecticide application to natural mixtures of the grass *Holcus mollis* and the herb *Galium saxatile* suggests that removal of the grass aphid, *Holcaphis holci,* increases the vigor of the grass to the point where *Galium* is substantially reduced in cover.[201b]

In a comparison of the first 2 years of secondary succession at two sites (one in southern England and one in Iowa), insect exclusion was found to have measurable effects on annual plant performance only at the English site, and then only in the first year.[33b] Conspicuous effects of insect exclusion were observed when malathion was applied to sheep-grazed, early successional plots in calcareous grassland;[111] *Medicago lupulina* flourished on the sprayed plots, probably as a result of the exclusion of *Sitona* weevils.[110a] Note, however, that *M. lupulina* also increases following small mammal exclosure,[134a] and insecticide treatment might have some deterrent effect on small mammal feeding.

Where positive responses to insect exclusion are observed, it is often not clear by what mechanism the plant population has benefited. For example, with soil-applied insecticides, it may be that the exclusion of insects such as collembola that graze on mycorrhizal fungi would lead to increased efficiency in water and nutrient uptake.[102a]

The early impression from insecticide studies is that while normal densities of insect herbivores may affect plant reproductive performance,[36,65,177] they often have little or no measurable impact on plant recruitment and have rarely been shown to have a marked influence on plant population dynamics or on community structure. Insecticide exclusion has never been used in an experimental attempt to reverse a successful case of biological weed control (see above) where it might be shown that, in addition to reducing the weed from its former abundance, the insect herbivores continue to regulate the plant population at stable low densities.

In short, it is clear that insects can affect plant performance in natural habitats, sometimes dramatically. Their effects on plant population dynamics, however, are typically subtle. There are very few examples where insect exclusion has had as great an impact as the erection of a fence to exclude vertebrate herbivores. We have yet to witness the publication of more than a handful of studies that demonstrate a profound role for insect herbivores in plant community dynamics.[37,111,176,183] There are just as many studies where these effects have been sought but not found.[78,91,220,222]

C. SUMMARY

Comparison of fencng and insecticide-exclusion experiments leaves little room for doubt that in most ecosystems, vertebrate herbivores have a substantially greater impact on plant population dynamics and community structure than invertebrates. Even in forest communities, where the main canopy defoliators tend to be invertebrates, vertebrates may still exert a controlling influence on plant dynamics through seed and seedling predation. Vertebrate herbivores may also exert important, indirect effects on potentially dominant, woody species by modifying the structure of the vegetation and hence the microclimate of the forest floor.

IV. CONCLUSIONS

A. THEORETICAL PREDICTIONS

Both vertebrate and invertebrate herbivores influence plant performance in essentially the same way, reducing growth and fecundity, delaying flowering, and increasing plant death rates. The fundamental differences between them are in the size of the individual "bites" they take (reflecting the large difference in their average body mass) and in their average population density per unit of food resource. Differences in feeding rate, metabolic rate, and mobility of the juvenile stages mean that the starvation tolerance and numerical responses of the two groups are also quite different.

The difference in diet breadth between vertebrates and invertebrates is important, because it means that the more polyphagous vertebrate herbivores occur at extremely high population densities relative to the abundance of some of their highly preferred but less common food plants. These plants suffer high rates of depletion (they are decreasers). For the typically monophagous invertebrate herbivores, there is no buffer provided by an abundant but relatively low-quality food that might allow their populations to be maintained at high densities while preferred plant species are depleted. Where decreased plant density has been detected under invertebrate herbivory, the animals have been relatively large, polyphagous, mobile species like grasshoppers,[206] the "honorary vertebrates" of Section III.C.

Particular kinds of herbivores tend to be associated with particular kinds of plant damage, at least when we think of pest species: mollusks with seedling losses, browsing mammals with losses of saplings, insects with seed destruction, etc. These kinds of classification rarely bear close scrutiny, however. Associations that look appealing at first glance (e.g., grasses and ungulates) do not stand up to experimental study (witness the important role of phloem-feeding, root-feeding, and stem-boring insects in grass growth).

The question of the relative importance of food regulaton for vertebrate and invertebrate herbivore populations must remain open. There are far too few cases in which population regulation is understood clearly, for either kind of herbivore. There are examples of food-limited vertebrates (voles,[104] wildebeest,[225] and reindeer[227]) and food-limited invertebrates (cinnabar moth,[88] grasshoppers,[205] and knopper gall wasp[68]). The most convincing demonstrations of natural enemy regulation have come from the resurgence of herbivorous invertebrates following the application of broad-spectrum insecticides; species have become pests since the insecticide killed their predators and parasitoids.[86] The evidence for natural enemy regulation of vertebrate herbivore populations is much weaker, often anecdotal,[56] and rarely subject to critical experimental tests in the field (but see work by Henttonen).[125] This imbalance in the weight of evidence suggests that natural enemy regulation may be more prevalent among invertebrate herbivores, and therefore that food regulation may be more common among vertebrate herbivores. It is important to bear in mind that population regulation by low food quality, rather than by an absolute shortage of food plants, may well be widespread among herbivorous insects, but that this is not likely to be associated with a reciprocal impact of insect feeding on plant population dynamics.[64]

B. EXPERIMENTAL RESULTS

Exclusion experiments, using fences to keep out vertebrates, leave no doubt as to the profound effect of vertebrate herbivory on vegetation structure and successional dynamics. Insect exclusion experiments are in their infancy, but results to date suggest that the effects of insect herbivores are markedly less pronounced in terms of vegetation stature, species composition, and relative plant abundance.

In strongly pulse-driven systems (e.g., in areas of low and unpredictable rainfall) where the dominant plant species may be long-lived and recruit from seed infrequently, and under very special circumstances, long-term field experiments may be doomed to show a lack of any significant impact for herbivores of any kind. Only if, by good luck or good judgment, the system is studied during one of these rare bouts of plant recruitment, will it be possible to assess the relative importance of vertebrate and invertebrate herbivory in plant dynamics.

At this stage, it would be premature to conclude that insect herbivores are unimportant in plant community dynamics. What has been shown is that vertebrate herbivores tend to be *more* important. It should be emphasized, however, that the distinction is not between vertebrates and invertebrates in any strict, taxonomic sense, but rather between small, sedentary, typically monophagous herbivores on the one hand and large, mobile, highly polyphagous herbivores on the other.

C. FUTURE STUDIES

The most important experiments that need to be performed involve fencing, insecticide, and molluscicide treatments, carried out in replicated, factorial combinations in a wide range of different plant communities. In each case, information should be gathered on the performance of individual plants, all the way from seed to adulthood. Since many of the effects of herbivory are likely to be felt through altered competitive relationships with neighboring plants,[61,107] the detailed studies of the effects of herbivory at the community level will inevitably involve a study of plant competition. Thus, all the extra difficulties attendant on competition studies will need to be confronted.[19,170]

It is vital that we learn how herbivore populations are regulated under field conditions. For the moment we can conclude little more than that in most communities, most of the time, vertebrate herbivores are more likely to be food-limited and hence to have the more pronounced impact on plant community dynamics. Where invertebrate herbivores are food-limited, it is more likely that access to high-quality food is the key factor, rather than the abundance of the host plant, and that the effects of herbivore feeding on plant population dynamics are consequently rather low. For both vertebrate and invertebrate herbivores, we should expect asymmetries in plant-herbivore interactions to be widespread, and that most of the asymmetries are such that plants have more effect on the population dynamics of herbivores than herbivores have on the population dynamics of plants.

REFERENCES

1. **Abbot, I. and Van Heurck, P.,** Comparison of insects and vertebrates as removers of seed and fruit in a Western Australian forest, *Aust. J. Ecol.,* 10, 165, 1985.
2. **Andersen, A. N.,** Effects of seed predation by ants on seedling densities at a woodland site in SE Australia, *Oikos,* 48, 171, 1987.
3. **Anderson, M. and Jonasson, S.,** Rodent cycles in relation to food resources on an alpine health, *Oikos,* 46, 93, 1985.
4. **Andow, D. A.,** Microsite of the green rice leafhopper *Nephrotettix cincticeps* (Homoptera: Cicadellidae), on rice: plant nitrogen and leafhopper density. *Res. Popul. Ecol. (Kyoto),* 26, 313, 1984.
5. **Annecke, D. P., Karny, M., and Burger, W. A.,** Improved biological control of the prickly pear, *Opuntia megacantha* Salm-Dyck, in South Africa through the use of an insecticide, *Phytophylactica,* 1, 9, 1969.
6. **Archer, A. and Detling, J. K.,** The effects of defoliation and competition on regrowth of tillers of two North American mixed grass prairie graminoids, *Oikos,* 43, 351, 1984.
7. **Askew, R. R.,** On the biology of the inhabitants of oak galls of the Cynipidae (Hymenoptera) in Britain, *Trans. Br. Soc. Entomol.,* 14, 237, 1961.
8. **Augspurger, C. K.,** Seedling survival of tropical tree species: interactions of dispersal distance, light-gaps and pathogens, *Ecology,* 65, 1705, 1984.
9. **Auld, T. D.,** Seed predation in native legumes of southeastern Australia, *Aust. J. Ecol.,* 8, 367, 1983.
10. **Bakker, J. P.,** The impact of grazing on plant communities, plant populations and soil conditions on salt marshes, *Vegetatio,* 62, 391, 1985.
11. **Bakker, J. P., de Bie, S., Dallinga, J. H., Tjaden, P., and de Vries, Y.,** Sheep-grazing as a management tool for heathland conservation and regeneration in the Netherlands, *J. Appl. Ecol.,* 20, 541, 1983.

12. **Ballardie, R. T. and Whelan, R. J.,** Masting, seed dispersal and seed predation in the cycad *Macrozamia communis, Oecologia,* 70, 100, 1986.
13. **Batzli, G. O.,** Nutritional ecology of the California vole: effects of food quality on reproduction, *Ecology,* 67, 406, 1986.
14. **Bazely, D. R. and Jefferies, R. L.,** Changes in the composition and standing crop of salt marsh communities in response to the removal of a grazer, *J. Ecol.,* 74, 693, 1986.
15. **Becker, P., Lee, L. L., Rothman, E. D., and Hamilton, W. D.,** Seed predation and the coexistence of tree species: Hubbell's models revisited, *Oikos,* 44, 382, 1985.
16. **Bell, R. H. V.,** The use of the herb layer by grazing ungulates in the Serengeti, in *Animal Populations in Relation to their Food Resources,* Watson, A., Ed., Blackwell Scientific, London, 1970, 111.
17. **Belsky, A. J.,** Population and community processes in a mosaic grassland in the Serengeti, Tanzania, *J. Ecol.,* 74, 841, 1986.
18. **Belsky, A. J.,** Does herbivory benefit plants? A review of the evidence, *Am. Nat.,* 127, 870, 1986.
19. **Bender, E. A., Case, T. J., and Gilpin, M. E.,** Perturbation experiments in community ecology: theory and practice, *Ecology,* 65, 1, 1984.
20. **Benkman, C. W., Balda, R. P., and Smith, C. C.,** Adaptations for seed dispersal and the compromises due to seed predation in limber pine, *Ecology,* 65, 632, 1984.
21. **Bennett, A. and Krebs, J.,** Seed dispersal by ants, *Trends Ecol. Evol.,* 2, 291, 1987.
22. **Bergquist, A. M. and Carpenter, S. R.,** Limnetic herbivory: effects on phytoplankton populations and primary production, *Ecology,* 67, 1351, 1986.
23. **Bernays, E. A. and Lewis, A. C.,** The effect of wilting on palatability of plants to *Schistocerca gregaria,* the desert locust, *Oecologia,* 70, 132, 1986.
24. **Berryman, A. A., Dennis, B., Raffa, K. F., and Stenseth, N. C.,** Evolution of optimal group attack, with particular reference to bark beetles (Coleoptera: Scolytidae), *Ecology,* 66, 898, 1985.
25. **Berryman, A. A., Stenseth, N. C., and Isaev, A. S.,** Natural regulation of herbivorous forest insect populations, *Oecologia,* in press.
26. **Bishop, G. F. and Davy, A. J.,** Significance of rabbits for the population regulation of *Hieracium pilosella* in Breckland, *J. Ecol.,* 72, 273, 1984.
27. **Blaney, W. M. and Simmonds, M. S. J.,** Food selection by locusts: the role of learning in rejection behaviour, *Entomol. Exp. Appl.,* 39, 273, 1985.
28. **Borowicz, V. A. and Juliano, S. A.,** Inverse density-dependent parasitism of *Cornus amomum* fruit by *Rhagoletis cornivora, Ecology,* 67, 639, 1986.
29. **Boscher, J.,** Modified reproduction strategy of leek *Allium porrum* in response to a phytophagous insect, *Acrolepiopsis assectella, Oikos,* 33, 451, 1979.
30. **Braithwaite, R. W., Dudzinski, M. L., Ridpath, M. G., and Parker, B. S.,** The impact of water buffalo on the monsoon forest ecosystem in Kakadu National Park, *Aust. J. Ecol.,* 9, 309, 1984.
31. **Breen, J. P. and Teestes, G. L.,** Relationships of yellow sugarcane aphid (Homoptera: Aphididae) density to sorghum damage, *J. Econ. Entomol.,* 79, 1106, 1986.
32. **Bronmark, C.,** Interactions between macrophytes, epiphytes and herbivores: an experimental approach, *Oikos,* 45, 26, 1985.
33. **Brookman, S. S.,** Evidence for selective fruit production in *Asclepias, Evolution,* 38, 72, 1984.

33a. **Brown, V. C. and McNeill, S.,** personal communication.

33b. **Brown, V. K. and Hendrix, S.,** personal communication.

34. **Brown, J. H. and Maurer, B. A.,** Body size, ecological dominance and Cope's rule, *Nature (London),* 324, 248, 1986.
35. **Brown, J. H., Reichman, O. J., and Davidson, D. W.,** Granivory in desert ecosystems, *Adv. Ecol. Res.,* 10, 201, 1979.
36. **Brown, V. K.,** The phytophagous insect community and its impact on early successional habitats, in *Proc. 5th Int. Symp. Insect-Plant Relationships,* Visser, J. H. and Minks, A. K., Eds., Pudoc, Wageningen, The Netherlands, 1982, 205.
37. **Brown, V. K., Gange, A. C., Evans, I. M., and Storr, A. L.,** The effect of insect herbivory on the growth and reproduction of two annual *Vicia* species at different stages in plant succession, *J. Ecol.,* 75, 1173, 1987.
38. **Bryant, J. P. and Kuropat, P. J.,** Selection of winter forage by subartic browsing vertebrates: the role of plant chemistry, *Annu. Rev. Ecol. Syst.,* 11, 261, 1980.
39. **Bultman, T. L. and Faeth, S. H.,** Selective oviposition by a leaf miner in response to temporal variation in abscission, *Oecologia,* 69, 117, 1986.
40. **Burgess, R. S. L. and Ennos, R. A.,** Selective grazing of acyanogenic white clover: variation in behaviour among populations of the slug *Deroceras reticulatum, Oecologia,* 73, 432, 1987.
41. **Cargill, S. M. and Jefferies, R. J.,** The effects of grazing by lesser snow geese on the vegetation of a sub-arctic salt marsh, *J. Appl. Ecol.,* 21, 669, 1984.
42. **Carroll, C. R. and Risch, S. J.,** The dy-

namics of seed harvesting in early successional communities by a tropical ant, *Solenopsis geminata, Oecologia,* 61, 388, 1984.

43. **Carter, W.,** *Insects in Relation to Plant Disease,* 2nd ed., John Wiley & Sons, New York, 1973.
44. **Cates, R. G.,** The interface between slugs and wild ginger: some evolutionary aspects, *Ecology,* 56, 391, 1975.
45. **Caughley, G.,** Eruption of ungulate populations, with emphasis on Himalayan thar in New Zealand, *Ecology,* 51, 53, 1970.
46. **Cavers, P. B. and Steel, M. G.,** Patterns of change in seed weight over time on individual plants, *Am. Nat.,* 124, 324, 1984.
47. **Chapman, R. F.,** The chemical inhibition of feeding by phytophagous insects: a review, *Bull. Entomol. Res.,* 64, 339, 1974.
48. **Chettleburgh, M. R.,** Observations on the collection and burial of acorns by jays in Hainault Forest, *Br. Birds,* 45, 359, 1952.
49. **Cideciyan, M. A. and Malloch, A. J. C.,** Effects of seed size on the germination, growth and competitive ability of *Rumex crispus* and *Rumex obtusifolius, J. Ecol.,* 70, 227, 1982.
50. **Claridge, M. F. and Wilson, M. R.,** Host plant associations, diversity and species-area relationships of mesophyll-feeding leafhoppers of trees and shrubs in Britain, *Ecol. Entomol.,* 6, 217, 1981.
51. **Clark, D. A. and Clark, D. B.,** Spacing dynamics of a tropical rain forest tree: evaluation of the Janzen-Connell model, *Am. Nat.,* 124, 769, 1984.
52. **Clements, R. O. and Henderson, I. F.,** Insects as a cause of botanical change in swards, *J. Br. Grassl. Soc. Occas. Symp.,* 10, 157, 1979.
53. **Coe, M. J., Bourn, D., and Swingland, I. R.,** The biomass, production and carrying capacity of giant tortoises on Aldabra, *Philos. Trans R. Soc. London Ser. B,* 286, 163, 1979.
54. **Coley, P. D.,** Effects of leaf age and plant life history patterns on herbivory, *Nature (London),* 284, 545, 1980.
55. **Coley, P. D., Bryant, J. P., and Chapin, F. S.,** Resource availability and plant antiherbivore defense, *Science,* 230, 895, 1985.
56. **Colinvaux, P. A.,** *Introduction to Ecology,* John Wiley & Sons, New York, 1973, 397.
57. **Collins, S. L. and Uno, G. E.,** Seed predation, seed dispersal and disturbance in grasslands: a comment, *Am. Nat.,* 125, 866, 1985.
58. **Connell, J. H.,** On the role of natural enemies in preventing competitive exclusion in some marine animals and in rain forests, in *Population Dynamics,* Anderson, R. M., Turner, B. D., and Taylor, L. R., Eds., Blackwell Scientific, Oxford, 1971, 141.
59. **Cooper, S. M. and Owen-Smith, N.,** Condensed tannins deter feeding by browsing ruminants in South African savanna, *Oecologia,* 67, 142, 1985.
60. **Cottam, D. A.,** Frequency-dependent grazing by slugs and grasshoppers, *J. Ecol.,* 73, 925, 1985.
61. **Cottam, D. A., Whittaker, J. B., and Malloch, A. J. C.,** The effects of chrysomelid beetle grazing and plant competition on the growth of *Rumex obtusifolius, Oecologia,* 70, 452, 1986.
62. **Cox, C. S. and McEvoy, P. B.,** Effect of summer moisture stress on the capacity of tansy ragwort *(Senecio jacobaea)* to compensate for defoliation by cinnabar moth *(Tyria jacobaeae), J. Appl. Ecol.,* 20, 225, 1983.
63. **Crawley, M. J.,** The numerical responses of insect predators to changes in prey density, *J. Anim. Ecol.,* 44, 877, 1975.
64. **Crawley, M. J.,** *Herbivory. The Dynamics of Animal-Plant Interactions,* Blackwell Scientific, Oxford, 1983.
65. **Crawley, M. J.,** Reduction of oak fecundity by low density herbivore populations, *Nature (London),* 314, 163, 1985.
66. **Crawley, M. J.,** The population biology of invaders, *Philos. Trans. R. Soc. London Ser. B,* 314, 711, 1986.
67. **Crawley, M. J.,** Benevolent herbivores?, *Trends Ecol. Evol.,* 2, 167, 1987.
68. **Crawley, M. J.,** The effects of insect herbivores on the growth and reproductive performance of English oak, in *Insects-Plants,* Labeyrie, V., Fabres, G., and Lachaise, D., Eds., Junk, The Hague, 1987, 307.
69. **Crawley, M. J.,** What makes a community invasible?, in *Colonization, Succession and Stability,* Gray, A. J., Crawley, M. J., and Edwards, P. J., Eds., Blackwell Scientific, Oxford, 1987, 429.
70. **Crawley, M. J.,** The effect of grazing by rabbits on individual size variation in winter wheat, *J. Appl. Ecol.,* submitted.
71. **Crawley, M. J.,** Rabbit grazing, plant competition and seedling recruitment in acid grassland, *J. Ecol.,* submitted.
72. **Crawley, M. J.,** Rabbits as pests of winter wheat, *Proc. Zool. Soc. London,* in press.
73. **Crawley, M. J.,** Herbivores and plant population dynamics, in *Plant Population Biology,* Davy, A. J., Hutchings, M. J., and Watkinson, A. R., Eds., Blackwell Scientific, Oxford, 1988.
74. **Crawley, M. J.,** Insect herbivores and plant population dynamics, *Annu. Rev. Entomol.,* in press.
75. **Crawley, M. J. and Brown, R. A.,** The im-

pact of rabbit grazing on the yield of winter wheat, *J. Appl. Ecol.,* submitted.
76. **Crawley, M. J. and Brown, R. A.,** The effects of intense rabbit grazing on growth and dry matter partitioning in winter wheat, *J. Appl. Ecol.,* submitted.
77. **Crawley, M. J. and Nachapong, M.,** The establishment of seedlings from primary and regrowth seeds of ragwort *(Senecio jacobaea), J. Ecol.,* 73, 255, 1985.
78. **Crawley, M. J., Islam, Z., Nachapong, M., Ohgushi, M., Pattrasudhi, R., Sheppard, A. W., Slater, A. F. G., and Wilcox, A.,** The population dynamics of cinnabar moth in mesic grasslands, *J. Anim. Ecol.,* submitted.
78a. **Crawley, M. J.,** unpublished results.
79. **Cubit, J. D.,** Herbivory and the seasonal abundance of algae on a high intertidal rocky shore, *Ecology,* 65, 1904, 1984.
80. **Damuth, J.,** Population density and body size in mammals, *Nature (London),* 290, 699, 1981.
81. **Darwin, C.,** *The Origin of Species,* John Murray, London, 1859.
82. **Davidar, P. D. and Morton, E. S.,** The relationship between fruit crop sizes and fruit removal rates by birds, *Ecology,* 67, 262, 1986.
83. **Davidson, D. W., Samson, D. A., and Inouye, R. S.,** Granivory in the Chihuahuan Desert: interactions within and between trophic levels, *Ecology,* 66, 486, 1985.
84. **De Leeuw, J. and Bakker, J. P.,** Sheepgrazing with different foraging efficiencies in a Dutch mixed grassland, *J. Appl. Ecol.,* 23, 781, 1986.
85. **De Steven, D. and Putz, F. E.,** Impact of mammals on early recruitment of a tropical canopy tree, *Dipteryx panamensis,* in Panama, *Oikos,* 43, 207, 1984.
86. **Debach, P.,** *Biological Control by Natural Enemies,* Cambridge University Press, Cambridge, 1974.
87. **Demment, M. W. and Van Soest, P. J.,** A nutritional explanation for body-size patterns of ruminant and nonruminant herbivores, *Am. Nat.,* 125, 641, 1985.
88. **Dempster, J. P.,** The natural control of populations of butterflies and moths, *Biol. Rev.,* 58, 461, 1983.
89. **Dewey, J. E.,** Western spruce budworm impact on Douglas-fir cone production, *Gen. Tech. Rep. Intermt. Res. Stn.,* 203, 243, 1986.
90. **Dirzo, R. and Harper, J. L.,** Experimental studies on slug-plant interaction. IV. The performance of cyanogenic and acyanogenic morphs of *Trifolium repens* in the field, *J. Ecol.,* 70, 119, 1982.
91. **Duggan, A. E.,** The Population Biology of Orange Tip Butterfly and its Host Plant, *Cardamine Pratensis,* Ph.D. thesis, University of London, England, 1988.
92. **Duggins, D. O. and Dethier, M. N.,** Experimental studies of herbivory and algal competition in a low intertidal habitat, *Oecologia,* 67, 183, 1985.
93. **Dunn, J. P., Kimmerer, T. W., and Nordin, G. L.,** The role of host tree condition in attack of white oaks by the twolined chestnut borer, *Agrilus bilineatus* (Weber) (Coleoptera: Buprestidae), *Oecologia,* 70, 596, 1986.
94. **Edmunds, G. F. and Alstad, D. N.,** Response of black pineleaf scales to host plant variability, in *Insect Life History Patterns and Habitat and Geographic Variation,* Denno, R. and Dingle, H., Eds., Springer-Verlag, New York, 1982, 29.
95. **Edwards, P. J.,** Herbivores in succession, in *Colonization, Succession and Stability,* Gray, A. J., Crawley, M. J., and Edwards, P. J., Eds., Blackwell Scientific, Oxford, 1987.
96. **Ellison, L.,** Influence of grazing on plant succession of rangelands, *Bot. Rev.,* 26, 1, 1960.
97. **Ellner, S. and Shmida, A.,** Why are adaptations for long-range seed dispersal rare in desert plants?, *Oecologia,* 51, 133, 1981.
98. **Elmqvist, T., Danell, K., Ericson, L., and Salomonson, A.,** Flowering and seed set in a boreal willow: the influence of vole herbivory, *Ecology,* in press.
99. **Elton, C.,** *The Ecology of Invasions by Animals and Plants,* Methuen, London, 1958.
100. **Faeth, S. H.,** Indirect interactions between temporally separated herbivores mediated by the host plant, *Ecology,* 67, 479, 1986.
101. **Faeth, S. H.,** Community structure and folivorous insect outbreaks: the roles of vertical and horizontal interactions, in *Insect Outbreaks,* Borbosat, P. and Shultz, J. C., Eds., Academic Press, New York, in press.
102. **Fagerstrom, T., Larsson, S., and Tenow, O.,** On optimal defence in plants, *Functional Ecol.,* 1, 73, 1987.
102a. **Fitter, A.,** personal communication.
103. **Flueckiger, W. and Braun, S.,** Effect of air pollutants on insects and hostplant/insect relationships, in *How are the Effects of Air Pollutants on Agricultural Crops Influenced by the Interaction with Other Limiting Factors,* Commission of the European Communities, Brussels, 1986, 79.
104. **Ford, R. G. and Pitelka, F. A.,** Resource limitation in populations of the California vole, *Ecology,* 65, 122, 1984.
104a. **Forrester, G.,** personal communication.
104b. **Forrester, G.,** unpublished data.
105. **Foster, S. A. and Janson, C. H.,** The relationship between seed size and establishment conditions in tropical woody plants, *Ecology,* 66, 773, 1985.

106. **Fowler, H. G. and Saes, N. B.,** Dependence of the activity of grazing cattle on foraging grass-cutting ants (*Atta* spp.) in the southern Neotropics, *J. Appl. Entomol.,* 101, 154, 1986.
107. **Fowler, N. L. and Rausher, M. D.,** Joint effects of competitors and herbivores on growth and reproduction in *Aristolochia reticulata, Ecology,* 66, 1580, 1985.
108. **Fox, L. R. and Morrow, P. A.,** Estimates of damage by herbivorous insects on Eucalyptus trees, *Aust. J. Ecol.,* 8, 139, 1983.
109. **Fye, R. E.,** Damage to vegetable and forage seedlings by overwintering *Lygus hesperus* (Heteroptera: Miridae) adults, *J. Econ. Entomol.,* 77, 1141, 1984.
110. **Galford, J. R.,** Primary infestation of sprouting chestnut, red, and white oak acorns by *Valentinia glandulella* (Lepidoptera: Blastobasidae), *Entomol. News,* 97, 109, 1986.
110a. **Gibson, C. W. D.,** personal communication.
111. **Gibson, C. W. D., Brown, V. K., and Jepsen, M.,** Relationships between the effects of insect herbivory and sheep grazing on seasonal changes in an early successional plant community, *Oecologia,* 71, 245, 1987.
112. **Goldberg, D. E.,** Effects of soil pH, competition, and seed predation on the distributions of two tree species, *Ecology,* 66, 503, 1985.
113. **Grant, S. A., Bolton, G. R., and Torvell, L.,** The responses of blanket bog vegetation to controlled grazing by hill sheep, *J. Appl. Ecol.,* 22, 739, 1985.
114. **Griffin, J. R.,** Oak regeneration in the Upper Carmel Valley, California, *Ecology,* 52, 862, 1971.
115. **Hainsworth, F. R., Wolf, L. L., and Mercier, T.,** Pollination and pre-dispersal seed predation: net effects on reproduction and inflorescence characteristics in *Ipomopsis aggregata, Oecologia,* 63, 405, 1984.
116. **Hairston, N. G., Smith, F. E., and Slobodkin, L. B.,** Community structure, population control and competition, *Am. Nat.,* 94, 421, 1960.
117. **Halse, S. A. and Trevenen, H. J.,** Damage to medic pastures by skylarks in North-western Iraq, *J. Appl. Ecol.,* 22, 337, 1985.
118. **Hannon, S. J., Mumme, R. L., Koenig, W. D., Spon, S., and Pitelka, F. A.,** Poor acorn crop, dominance and decline in numbers of acorn woodpeckers, *J. Anim. Ecol.,* 56, 197, 1987.
119. **Harper, J. L.,** *Population Biology of Plants,* Academic Press, London, 1977.
120. **Harrold, C. and Reed, D. C.,** Food availability, sea urchin grazing, and kelp forest community structure, *Ecology,* 66, 1160, 1985.
121. **Hay, M.,** Associational plant defences and the maintenance of species diversity: turning competitors into accomplices, *Am. Nat.,* 128, 617, 1986.
122. **Hay, M. E. and Taylor, P. R.,** Competition between herbivorous fishes and urchins on Caribbean reefs, *Oecologia,* 65, 591, 1985.
123. **Heichel, G. H. and Turner, N. C.,** Branch growth and leaf numbers of red maple (*Acer rubrum* L.) and red oak (*Quercus rubra* L.): response to defoliation, *Oecologia,* 62, 1, 1984.
123a. **Hendrix, S. D.,** personal communication.
124. **Hendrix, S. D.,** Relations of *Heracleum lanatum* to floral herbivory by *Depressaria pastinacella, Ecology,* 65, 191, 1984.
125. **Henttonen, H.,** Predation causing extended low densities in microtine cycles: further evidence from shrew dynamics, *Oikos,* 45, 156, 1985.
126. **Hobbs, R. J.,** Harvester ant foraging and plant species distribution in annual grassland, *Oecologia,* 67, 519, 1985.
127. **Holsten, E. H.,** Factors of susceptibility in spruce beetle attack on white spruce in Alaska, *J. Entomol. Soc. B.C.,* 81, 39, 1984.
128. **Hornfedt, B., Lofgren, O., and Carlsson, B.-G.,** Cycles in voles and small game in relation to variations in plant production indices in Northern Sweden, *Oecologia,* 68, 496, 1986.
129. **Horvitz, C. C. and Schemske, D. W.,** Antnest soil and seedling growth in a neotropical ant-dispersed herb, *Oecologia,* 70, 318, 1986.
130. **Hosking, G. P. and Hutcheson, J. A.,** Hard beech *(Nothofagus truncata)* decline on the Mamaku Plateau, North Island, New Zealand, *N.Z. J. Bot.,* 24, 263, 1986.
131. **Howarth, F. G.,** Impacts of alien land arthropods and mollusks on native plants and animals in Hawaii, in *Hawaii's Terrestrial Ecosystems: Preservation and Management,* Stone C. P. and Scott, J. M., Eds., University of Hawaii, Honolulu, 1985, 149.
132. **Howe, H. F. and Westley, L. C.,** Ecology of pollination and seed dispersal, in *Plant Ecology,* Crawley, M. J., Ed., Blackwell Scientific, Oxford, 1986, 185.
133. **Hudson, P. J.,** The effect of a parasitic nematode on the breeding production of red grouse, *J. Anim. Ecol.,* 55, 85, 1986.
134. **Hughes, R. E. and Dale, J.,** Trends in montane grasslands in Snowdonia, expressed in terms of 'relative entropy', *Nature (London),* 225, 756, 1970.
134a. **Hulme, P.,** personal communication.
135. **Humphreys, W. F.,** Production and respiration in animal populations, *J. Anim. Ecol.,* 48, 427, 1979.
136. **Huntly, N. J.,** Influence of refuging consumers (pikas, *Ochotona princeps*) on subalpine meadow vegetation, *Ecology,* in press.
137. **Ingham, R. E. and Detling, J. K.,** Plant-

herbivore interaction in a North American mixed-grass prairie. III. Soil nematode populations and root biomass on *Cynomys ludovicianus* colonies and adjacent uncolonized areas, *Oecologia,* 63, 307, 1984.
138. **Inouye, D. W.,** The consequences of herbivory: a mixed blessing for *Jurinea mollis* (Asteraceae), *Oikos,* 39, 269, 1982.
139. **Islam, Z. and Crawley, M. J.,** Compensation and regrowth in ragwort *(Senecio jacobaea)* attacked by cinnabar moth *(Tyria jacobaeae), J. Ecol.,* 71, 829, 1983.
140. **Janzen, D. H.,** Herbivores and the number of tree species in tropical forests, *Am. Nat.,* 104, 501, 1970.
141. **Janzen, D. H.,** Behaviour of *Hymenaea coubaril* when its predispersal seed predator is absent, *Science,* 189, 145, 1975.
142. **Janzen, D. H.,** New horizons in the biology of plant defences, in *Herbivores: Their Interaction with Secondary Plant Metabolites,* Rosenthal, G. A. and Janzen, D. H., Eds., Academic Press, New York, 1979, 331.
143. **Janzen, D. H.,** Patterns of herbivory in a tropical deciduous forest, *Biotropica,* 13, 271, 1981.
144. **Janzen, D. H.,** *Spondias mombin* is culturally deprived in megafauna-free forest, *J. Trop. Ecol.,* 1, 131, 1985.
145. **Janzen, D. H.,** Mice, big mammals, and seeds: it matters who defecates what where, in *Frugivores and Seed Dispersal,* Estrada, A. and Fleming, T. H., Eds., Junk, The Hague, 1986, 251.
146. **Janzen, D. H., Ryan, C. A., Liener, I. E., and Pearce, G.,** Potentially defensive proteins in mature seeds of 59 species of tropical Leguminosae, *J. Chem. Ecol.,* 12, 1469, 1986.
147. **Jara, H. F. and Moreno, C. A.,** Herbivory and structure in a midlittoral rocky community: a case in southern Chile, *Ecology,* 65, 28, 1984.
148. **Jarvinen, A.,** Predation causing extended low densities in microtine cycles: implications from predation on hole-nesting passerines, *Oikos,* 45, 157, 1985.
149. **Jensen, T. S.,** Seed production and outbreaks of non-cyclic rodent populations in deciduous forests, *Oecologia,* 54, 184, 1982.
150. **Jensen, T. S. and Nielsen, O. F.,** Rodents as seed dispersers in a health-oak wood succession, *Oecologia,* 70, 214, 1986.
151. **Joenje, W.,** The significance of waterfowl grazing in the primary vegetation succession on embanked sandflats, *Vegetatio,* 62, 399, 1985.
152. **Jonasson, S., Bryant, J. P., Chapin, F. S., and Andersson, M.,** Plant phenols and nutrients in relation to variations in climate and rodent grazing, *Am. Nat.,* 128, 394, 1986.
153. **Jones, F. G. W. and Jones, M. G.,** *Pests of Field Crops,* Edward Arnold, London, 1964.
154. **Jones, M. G.,** Grassland management and its influence on the sward, *J. R. Agric. Soc.,* 94, 21, 1933.
155. **Kahn, D. M. and Cornell, H. V.,** Early leaf abscission and folivores: comments and considerations, *Am. Nat.,* 122, 428, 1983.
156. **Karban, R.,** Periodical cicada nymphs impose periodical oak tree wood accumulation, *Nature (London),* 287, 326, 1980.
157. **Karel, A. K. and Mghogho, R. M. K.,** Effects of insecticides and plant populations on the insect pests and yield of common bean (*Phaseolus vulgaris* L.), *J. Econ. Entomol.,* 78, 917, 1985.
158. **Keeley, J. E., Keeley, S. C., Swift, C. C., and Lee, J.,** Seed predation due to the Yucca-moth symbiosis, *Am. Midl. Nat.,* 112, 187, 1984.
159. **Kerner, A.,** *The Natural History of Plants. Their Forms, Growth, Reproduction and Distribution,* Blackie and Son, London, 1894.
160. **King, K. L. and Hutchinson, K. J.,** The effects of sheep grazing on invertebrate numbers and biomass in unfertilized natural pastures of the New England Tablelands (NSW), *Aust. J. Ecol.,* 8, 245, 1983.
161. **Kinkaid, W. B. and Cameron, G. N.,** Interactions of cotton rats with a patchy environment: dietary responses and habitat selection, *Ecology,* 66, 1769, 1985.
162. **Kinsman, S. and Platt, W. J.,** The impact of a herbivore upon *Mirabilis hirsuta,* a fugitive prairie plant, *Oecologia,* 65, 2, 1984.
163. **Kirkland, R. L. and Goeden, R. D.,** An insecticidal-check study of the biological control of puncturevine *(Tribulus terrestris)* by imported weevils, *Microlarinus lareynii* and *M. lypriformis* (Col.: Curculionidae), *Environ. Entomol.,* 7, 349, 1978.
164. **Kjellsson, G.,** Seed fate in a population of *Carex pilulifera* L. I. Seed dispersal and ant-seed mutualism. II. Seed predation and its consequences for dispersal and seed bank, *Oecologia,* 67, 416, 1985.
165. **Klein, D. R.,** The problems of overpopulation of deer in North America, in *Problems in Management of Locally Abundant Wild Mammals,* Jewell, P. A. and Holt, S., Eds., Academic Press, New York, 1981, 119.
166. **Ko, J. H. and Morimoto, K.,** Loss of tree vigor and role of boring insects in red pine stands heavily infested by the pine needle gall midge in Korea, *Esakia,* 23, 151, 1985.
167. **Lamberti, G. A., Feminella, J. W., and Resh, V. H.,** Herbivory and intraspecific competition in a stream caddisfly population, *Oecologia,* 73, 75, 1987.
168. **Lance, D. R., Elkington, J. S., and**

Schwalbe, C. P., Feeding rhythms of gypsy moth larvae: effects of food quality during outbreaks, *Ecology,* 67 1650, 1986.

169. **Lange, R. T. and Graham, C. R.,** Rabbits and the failure of regeneration in Australian arid zone Acacia, *Aust. J. Ecol.,* 8, 377, 1983.
170. **Law, R. and Watkinson, A. R.,** Response-surface analysis of two-species competition: an experiment on *Phleum arenarium* and *Vulpia fasciculata, J. Ecol.,* 75, 871, 1987.
171. **Laws, R. M., Parker, I. S. C., and Johnstone, R. C. B.,** *Elephants and Their Habitat,* Clarendon Press, Oxford, 1975.
172. **Lawton, J. H. and MacGarvin, M.,** The organization of herbivore communities, in *Community Ecology: Pattern and Process,* Kikkawa, J. and Anderson, D. J., Eds., Blackwell Scientific, Oxford, 1986, 163.
173. **Lindroth, R. L. and Batzli, G. O.,** Inducible plant chemical defences: a cause of vole population cycles?, *J. Anim. Ecol.,* 55, 431, 1986.
174. **Lodge, G. M. and Whalley, R. D. B.,** The manipulation of species composition of natural pastures by grazing management on the northern slopes of New South Wales, *Aust. Rangeland J.,* 7, 6, 1985.
175. **Louda, S. M.,** Limitation of the recruitment of the shrub *Haplopappus squarrosus* (Asteraceae) by flower- and seed-feeding insects, *J. Ecol.,* 70, 43, 1982.
176. **Louda, S. M.,** Seed predation and seedling mortality in the recruitment of a shrub, *Haplopappus venetus* (Asteracea), along a climatic gradient, *Ecology,* 64, 5, 1983.
177. **Louda, S. M.,** Herbivore effect on stature, fruiting, and leaf dynamics of a native crucifer, *Ecology,* 65, 137, 1984.
178. **Louda, S. M.,** Insect herbivory in response to root-cutting and flooding stress on a native crucifer under field conditions, *Acta Oecol.,* 7, 37, 1986.
179. **Maddock, L.,** The 'migration' and grazing succession, in *Serengeti: Dynamics of an Ecosystem,* Sinclair, A. R. E. and Norton-Griffiths, M., Eds., University of Chicago Press, Chicago, 1979, 104.
180. **Manzur, M. I. and Courtney, S. P.,** Influence of insect damage in fruits of hawthorn on bird foraging and seed dispersal, *Oikos,* 43, 265, 1984.
181. **Matthews, N. J. and Flegg, J. J. M.,** Seeds, buds and bullfinches, in *Pests, Pathogens and Vegetation,* Thresh, J. M., Ed., Pitman, London, 1981, 375.
182. **McAuliffe, J. R.,** Herbivore-limited establishment of a Sonoran desert tree, *Cercidium microphyllum, Ecology,* 67, 276, 1986.
183. **McBrien, H., Harmsen, R., and Crowder, A.,** A case of insect grazing affecting plant succession, *Ecology,* 64, 1035, 1983.
184. **McNaughton, S. J.,** Grazing lawns: animals in herds, plant form and coevolution, *Am. Nat.,* 124, 863, 1984.
185. **McNeill, S. and Lawton, J. H.,** Annual production and respiration in animal populations, *Nature (London),* 225, 472, 1970.
186. **van der Meijden, E., de Jong, T. J., Klinkhamer, P. G. L., and Kooi, R. E.,** Temporal and spatial dynamics in populations of biennial plants, in *Structure and Functioning of Plant Populations,* Vol. 2, Haeck, J. and Woldendorp, J. W., Eds., North-Holland, Amsterdam, 1985, 91.
187. **Messier, F. and Crete, M.,** Moose-wolf dynamics and the natural regulation of moose populations, *Oecologia,* 65, 503, 1985.
188. **Miller, G. E., Hedin, A. F., and Ruth, D. S.,** Damage by two Douglas-fir cone and seed insects: correlation with cone crop size, *J. Entomol. Soc. B.C.,* 81, 46, 1984.
189. **Mills, J. N.,** Herbivores and early postfire succession in southern California chaparral, *Ecology,* 67, 1637, 1986.
190. **Miyashita, T.,** Estimation of the economic injury level in the rice leaf-roller, *Cnaphalocrocis medinalis* Guenee (Lepidoptera, Pyralidae). I. Relation between yield loss and injury of rice leaves at heading or in the grain filling period, *Jpn. J. Appl. Entomol. Zool.,* 29, 73, 1985.
191. **Morrill, W. L., Ditterline, R. L., and Winstead, C.,** Effects of *Lygus borealis* Kelton (Hemiptera: Miridae) and *Adelphocoris lineolatus* (Goeze) (Hemiptera: Miridae) feeding on sainfoin production, *J. Econ. Entomol.,* 77, 966, 1984.
192. **Morris, R. F., Ed.,** The dynamics of epidemic spruce budworm populations, *Mem. Entomol. Soc. Can.,* 31, 1, 1963.
193. **Morse, D. R., Lawton, J. H., Dodson, M. M., and Williamson, M. H.,** Fractal dimensions of vegetation and the distribution of arthropod body lengths, *Nature (London),* 314, 731, 1985.
194. **Morton, S. R.,** Granivory in arid regions: comparison of Australia with North and South America, *Ecology,* 66, 1859, 1985.
195. **Myers, J. H.,** Is the insect or the plant the driving force in the cinnabar moth-tansy ragwort system?, *Oecologia,* 48, 151, 1980.
196. **Nelson, D. M. and Johnson, C. D.,** Stabilizing selection on seed size in *Astragalus* (Leguminosae) due to differential predation and differential germination, *J. Kans. Entomol. Soc.,* 56, 169, 1983.
197. **Noy-Meir, I.,** Responses of vegetation to the abundance of mammalian herbivores, in *Problems in Management of Locally Abundant Wild Mammals,* Jewell, P. A. and Holt, S., Eds., Academic Press, New York, 1981, 233.
198. **O'Dowd, D. J. and Gill, A. M.,** Predator

satiation and site alteration following fire: mass reproduction of alpine ash *(Eucalyptus delegatensis)* in southeastern Australia, *Ecology,* 65, 1052, 1984.

199. **O'Dowd, D. J. and Hay, M. E.,** Mutualism between harvester ants and a desert ephemeral: see escape from rodents, *Ecology,* 61, 531, 1980.
200. **Ohmart, C. P.,** Is insect defoliation in eucalypt forests greater than that in other temperate forests?, *Aust. J. Ecol.,* 9, 413, 1984.
201. **Oksanen, L.,** Trophic exploitation and arctic phytomass patterns, *Am. Nat.,* 122, 45, 1983.

201a. **Packham, J.,** personal communication.

201b. **Packham, J.,** personal communication.

202. **Paige, K. N. and Whitham, T. G.,** Overcompensation in response to mammalian herbivory: the advantage of being eaten, *Am. Nat.,* 129, 419, 1987.
203. **Pantoja, A., Smith, C. M., and Robinson, J. F.,** Effects of the fall armyworm (Lepidoptera: Noctuidae) on rice yields, *J. Econ. Entomol.,* 79, 1324, 1986.
204. **Parker, M. A.,** Local food depletion and the foraging behaviour of a specialist grasshopper, *Hesperotettix viridis, Ecology,* 65, 824, 1984.
205. **Parker, M. A.,** Size dependent herbivore attack and the demography of an arid grassland shrub, *Ecology,* 66, 850, 1985.
206. **Parker, M. A. and Root, R. B.,** Insect herbivores limit habitat distribution of a native composite, *Machaeranthera canescens, Ecology,* 62, 1390, 1981.
207. **Peters, R. H.,** *The Ecological Implications of Body Size,* Cambridge University Press, Cambridge, 1983.
208. **Pigott, C. D.,** Selective damage to tree-seedlings by bank voles *(Clethrionomys glareolus), Oecologia,* 67, 367, 1985.
209. **Potts, G. R., Tapper, S. C., and Hudson, P. J.,** Population fluctuations in red grouse: analysis of bag records and a simulation model, *J. Anim. Ecol.,* 52, 21, 1984.
210. **Price, M. V. and Waser, N. M.,** Microhabitat use by heteromyid rodents: effects of artificial seed patches, *Ecology,* 66, 211, 1985.
211. **Pyke, D. A.,** Demographic responses of *Bromus tectorum* and seedlings of *Agropyron spicatum* to grazing by small mammals: occurrence and severity of grazing, *J. Ecol.,* 74, 739, 1986.
212. **Rainey, R. C.,** Wild plants in the ecology of the desert locust, in *Pests, Pathogens and Vegetation,* Thresh, J. M., Ed., Pitman, London, 1981, 327.
213. **Rawes, M.,** Further results of excluding sheep from high level grasslands in the north Pennines, *J. Ecol.,* 69, 651, 1981.
214. **Rhoades, D. F.,** Offensive-defensive interactions between herbivores and plants: their relevance in herbivore population dynamics and ecological theory, *Am. Nat.,* 125, 205, 1985.
215. **Rice, B. and Westoby, M.,** Evidence against the hypothesis that anti-dispersed seeds reach nutrient-enriched microsites, *Ecology,* 67, 1270, 1986.
216. **Rice, K. J.,** Interaction of disturbance patch size and herbivory in *Erodium* colonization, *Ecology,* 68, 113, 1987.
217. **Richards, J. H.,** Root growth response to defoliation in two *Agropyron* bunchgrasses: field observations with an improved root periscope, *Oecologia,* 64, 21, 1984.
218. **Rissing, S. W.,** Indirect effects of granivory by harvester ants: plant species composition and reproductive increase near ant nests, *Oecologia,* 68, 231, 1986.
219. **Roach, D. A.,** Timing of seed production and dispersal in *Geranium carolinianum:* effects on fitness, *Ecology,* 67, 572, 1986.
220. **Sheppard, A. W.,** Insect Herbivore Competition and the Population Dynamics of *Heracleum sphondylium,* L. (Umbelliferae), Ph.D. thesis, University of London, England, 1987.
221. **Short, J.,** The functional response of kangaroos, sheep and rabbits in an arid grazing system, *J. Appl. Ecol.,* 22, 435, 1985.
222. **Shure, D. J.,** Insecticide effects on early succession in an old field ecosystem, *Ecology,* 52, 271, 1971.
223. **Silvertown, J. W.,** The evolutionary ecology of mast seeding in trees, *Biol. J. Linn. Soc.,* 14, 235, 1980.
224. **Sinclair, A. R. E.,** The eruption of the ruminants, in *Serengeti: Dynamics of an Ecosystem,* Sinclair, A. R. E. and Norton-Griffiths, M., Eds., University of Chicago Press, Chicago, 1979, 82.
225. **Sinclair, A. R. E., Dublin, H., and Borner, M.,** Population regulation of Serengeti wildebeest: a test of the food hypothesis, *Oecologia,* 65, 266, 1985.
226. **Singer, M. C. and Ehrlich, P. R.,** Population dynamics of the checkerspot butterfly *Euphydryas editha* in *Population Ecology,* Halbach, U. and Jacobs, J., Eds., Fischer-Verlag, Stuttgart, 1979, 29.
227. **Skogland, T.,** The effect of density-dependent resource limitations on the demography of wild reindeer, *J. Anim. Ecol.,* 54, 359, 1985.
228. **Smallwood, P. D. and Peters, W. D.,** Grey squirrel food preferences: the effects of tannin and fat concentration, *Ecology,* 67, 168, 1986.
229. **Stapanian, M. A. and Smith, C. C.,** Density-dependent survival of scatterhoarded nuts: an experimental approach, *Ecology,* 65, 1387, 1984.

230. **Stephenson, A. G.,** Flower and fruit abortion: proximate causes and ultimate function, *Annu. Rev. Ecol. Syst.,* 12, 253, 1981.
231. **Stephenson, A. G. and Winsor, J. A.,** *Lotus corniculatus* regulates offspring quality through selective fruit abortion, *Evolution,* 40, 453, 1986.
232. **Stork, V. L.,** Examination of seed dispersal and survival in red oak, *Quercus rubra* (Fagacea), using metal-tagged acorns, *Ecology,* 65, 1020, 1984.
233. **Stuart, J. D.,** Hazard rating of lodgepole pine stands to mountain pine beetle outbreaks in central Oregon, *Can. J. For. Res.,* 14, 666, 1984.
234. **Summers, D. D. B.,** Bullfinch *(Pyrrhula pyrrhula)* damage in orchards in relation to woodland bud and seed feeding, in *Pests, Pathogens and Vegetation,* Thresh, J. M., Ed., Pitman, London, 1981, 385.
235. **Tansley, A. G.,** *The British Islands and their Vegetation,* Cambridge University Press, Cambridge, 1939.
236. **Taylor, L. R.,** Insect migration, flight periodicity and the boundary layer, *J. Anim. Ecol.,* 43, 225, 1974.
237. **Thomas, A. S.,** Changes in vegetation since the advent of myxomatosis, *J. Ecol.,* 48, 287, 1960.
238. **Thomas, C. D.,** Butterfly larvae reduce host plant survival in vicinity of alternative host species, *Oecologia,* 70, 113, 1986.
239. **Thompson, J. N.,** Variation among individual seed masses in *Lomatium grayi* (Umbelliferae) under controlled conditions: magnitude and partitioning of the variance, *Ecology,* 65, 626, 1984.
240. **Thompson, J. N.,** Postdispersal seed predation in *Lomatium* spp. (Umbelliferae): variation among individuals and species, *Ecology,* 66, 1608, 1985.
241. **Townsend, L. H. and Sedlacek, J. D.,** Damage to corn caused by *Euschistus servus, E. variolarius,* and *Acrosternum hilare* (Heteroptera: Pentatomidae) under greenhouse conditions, *J. Econ. Entomol.,* 79, 1254, 1986.
242. **Varley, G. C. and Gradwell, G. R.,** Population models for the winter moth, in *Insect Abundance,* Southwood, T. R. E., Ed., Blackwell Scientific, Oxford, 1968, 132.
243. **Wagner, F. and Ehrhardt, R.,** Untersuchungen am Stickanal der Graswanze *Miris dolobratus* L., der Urheberin der totalen Weissahrigkeit der Rotschwingels *(Festuca rubra), Z. Pflanzenkr. Pflanzenschutz,* 68, 615, 1961.
244. **Wainhouse, D. and Howell, R. S.,** Intraspecific variation in beech scale populations and in susceptibility of their host *Fagus sylvatica, Ecol. Entomol.,* 8, 351, 1983.
245. **Waller, D. M.,** Models of mast fruiting in trees, *J. Theor. Biol.,* 80, 223, 1979.
246. **Ward, P.,** Feeding ecology of the black-faced dioch *Quelea quelea* in Nigeria, *Ibis,* 107, 173, 1965.
247. **Waring, R. H. and Pitman, G. B.,** Modifying lodgepole pine stands to change susceptibility to mountain pine beetle attack, *Ecology,* 66, 889, 1985.
248. **Watt, A. D.,** The performance of the pine beauty moth on water-stressed lodgepole pine plants: a laboratory experiment, *Oecologia,* 70, 578, 1986.
249. **Watt, A. S.,** The effect of excluding rabbits from acidiphilous grassland in Breckland, *J. Ecol.,* 48, 601, 1960.
250. **Welch, D.,** Studies in the grazing of heather moorland in north-east Scotland. V. Trends in *Nardus stricta* and other unpalatable graminoids, *J. Appl. Ecol.,* 23, 1047, 1986.
251. **Westhoff, V. and Sykora, K. V.,** A study of the influence of desalination on the *Juncetum gerardii, Acta Bot. Neerl.,* 28, 505, 1979.
252. **Westoby, M.,** Does heavy grazing usually improve the food resource for grazers?, *Am. Nat.,* 126, 870, 1985.

252a. **Westoby, M.,** personal communication.

253. **Westoby, M., Noy-Meir, I., and Walker, B. H.,** Irreversible rangeland successions, *J. Arid Lands Res.,* in press.
254. **Whitham, T. G.,** Evolution of territoriality by herbivores in response to host plant defences, *Am. Zool.,* in press.
255. **van Wijngaarden, W.,** Elephants-Trees-Grass-Grazers: Relationships between Climate Soil, Vegetation and Large Herbivores in a Semi-Arid Savanna Ecosystem (Tsavo, Kenya), ITC Publ. No. 4, Wageningen, The Netherlands, 1985, 1.
256. **Woolfenden, G. E. and Fitzpatrick, J. W.,** *The Florida Scrub Jay: Demography of a Cooperative-Breeding Bird,* Princeton, NJ, 1984.
257. **Wulff, R. D.,** Seed size variation in *Desmodium paniculatum.* I. Factors affecting seed size. II. Effects on seedling growth and physiological performance. III. Effects on reproductive yield and competitive ability, *J. Ecol.,* 74, 87, 1986.
258. **Yair, A. and Danin, A.,** Spatial variations in vegetation as related to the soil moisture regime over an arid limestone hillside, northern Negev, Israel, *Oecologia,* 47, 83, 1980.
259. **Young, T. P.,** *Lobelia telekii* herbivory, mortality, and size at reproduction: variation with growth rate, *Ecology,* 66, 1879, 1985.
260. **Zakaria, S.,** personal communication.

3 Air Pollution and Insect Herbivores: Observed Interactions and Possible Mechanisms

J. Riemer
Institute of Plant Ecology
University of Copenhagen, Denmark

J. B. Whittaker
Biological Sciences
University of Lancaster, England

TABLE OF CONTENTS

I. INTRODUCTION

There is a growing amount of evidence of a connection between air pollution and changes in insect attacks on plants. It ranges from very unspecific observations by forest managers to advanced laboratory experiments including biochemical explanations of the observed changes. This article reviews the present knowledge and attempts to outline the general mechanisms that seem to be involved in changes in insect communities in a polluted environment. The main attention is given to effects of SO_2, O_3, and oxides of nitrogen, as these pollutants are the most important on a regional scale. Effects of fluorides on insects have been documented in a number of cases,[2,137,148,184,194] but the effects are usually restricted to small areas around point sources, and fluorides will only be treated in connection with other pollutants.

II. FIELD OBSERVATIONS

Observations of numerical changes in the insect fauna in the field are very difficult to interpret in relation to causal factors, because natural insect populations fluctuate widely and there is serious lack of baseline data on the extent of natural changes. This is especially true for agricultural pests, where field evaluations are hampered by changing practices of agriculture over the years. For some forest pests there are, however, long-term records available from which tentative judgments of long-term trends can be made.

Despite the difficulties of establishing causal relationships from field studies, a number of observations have been made (mostly in the heavily industrialized parts of Central Europe) and some appear to show not only numerical changes but also changes in the structure of the arthropod communities.

Changes in the vicinity of strongly polluting industries have been known for at least 150 years. Counts of larvae of *Epinotia tedella* on Norway spruce around a German iron foundry between 1832 and 1833 showed that this lepidopteran was seven times more abundant in the fume zone than in nonpolluted adjacent areas.[37] Studies of smoke-damaged stands of pine in Poland have shown increased attack by the lepidopterans *Exoteleia dodecella, Petrova resinella, Rhyacionia buoliana,* and *Lymantria monacha,* while the needle-feeding hymenopteran *Acantholyda nemoralis* seems to disappear from stands damaged SO_2.[169] In the Kölner Bay area in northern Germany, the numbers of *Dreyfusia nüsslini* were much higher on *Abies concolor* that showed simultaneous injuries caused by sulfur dioxide and fluorides than on undamaged trees. The opposite observation was made for woolly aphids on *Pinus griffithi* and *P. sylvestris,* whose attack was strongest on trees with no signs of necrosis.[11] Flückiger and Flückiger-Keller[54] report abnormally high densities of the bark-feeding coccid *Cryptococcus fagi* on beech in the polluted valley and hills around Basel and found a correlation between the concentration of amino acids in the sap and the number of coccids. Pfeffer[149] reports that the galling adelgids *Sacchiphantes viridis, S. abietis, Adelges laricis,* and *A. tardus* are very numerous in zones of heavy air pollution. He attributed their high tolerance of pollution to their wax-cover and protected position, which may limit the direct impact of pollutants. Increased incidence of adelgids has also been reported from smoke-damaged stands in Germany.[46] Field studies in and around industrial cities of the Kemorovo province in Russia have likewise shown an increase in insects with sucking or piercing lifestyles in polluted areas, and among these, species with a cryptic or semicryptic behavior predominate.[6] An increase in aphid numbers was observed around a coal-fired power plant in Pennsylvania, where a concomitant decrease in the numbers of honeybees (*Apis mellifera*) and a number of parasitic hymenopterans was registered compared to a nonpolluted reference area. The numbers of social wasps seemed unaffected.[89,90] English surveys of the bark-living fauna around Newcastle showed a general reduction in the diversity of insect herbivores on trunks of trees in polluted areas, whereas the density and diversity of predators seemed undisturbed. The reduction in herbivore species may partly be explained by the disappearence of lichens on the trunks.[67]

Reports of increased attack from bark beetles in pollution-affected areas are numerous. Field surveys in an area affected by brown-coal-burning power plants have shown increased attack on smoke-damaged trees by a number of species of bark beetles.[19] Similar reports come from Czechoslovakia where the bark beetles *Phaenops cyanea* and *Pissodes piniphilus* on pine, and *P. harcynia* and *P. scabricollis* on spruce have been reported in large numbers in smoke-damaged stands.[149] Studies of bark-beetle infestations in oxidant-injured stands in California have shown that almost all the trees weakened by oxidants were ultimately killed by the western pine beetle *Dendroctonus brevicornis* or the mountain pine beetle *D. ponderosae,*[136] and the productivity of the bark beetles has been shown to be higher on oxidant-injured trees.[40] Oxidant-damaged trees not yet attacked by bark beetles exhibited a reduced rate of flow, exudation pressure, and yield of oleoresin, and the rate of resin crystallization was increased. These symptoms suggest that the trees would become less resistant towards bark beetles.[33,178] On the other hand, bark beetles of the monophagous *Pityokteines* family have been reported to disappear from smoke-damaged stands[46,148] and Führer[64] has demonstrated that the heavy impact of mixed industrial pollution on Norway spruce limits the success of *Pityogenes chalcographus* by reducing egg production and by increasing the mortality of eggs and larvae.

While effects close to point sources have been known for many years, there has been a growing number of reports recently that indicate more widespread changes in insect herbivory are occurring over large areas. Relating these changes to air pollution is extremely difficult, but it is possible that they are connected to the implementation of the high stack policy, which has changed the air pollution from localized high levels around point sources to a much more widespread phenomenon affecting very large areas.[195] The reason for the increasing numbers of reports may, however, simply be a growing awareness of changes in insect herbivory and should therefore be interpreted with some caution. One of the recurring observations is an increase over large areas of insect species that have hitherto been regarded as harmless because of their relatively stable, low population levels.[5] Some examples are outbreaks of *Cephalcia falleni* on spruce in Southern Poland (1981/82), *Gilpinia pallida* on pine in Austria (1971 and 1982), and *Ptilophora plumigera* on maple in Germany (1980 to 1982).[63]

The lepidopterans *Exoteleia dodecella* and *Rhyacionia buoliana* are reported in mass outbreaks in weakened stands of *Pinus sylvestris* in many parts of Europe.[46,149,168,172] Other lepidopterans that seem to become increasingly important in polluted areas are *Laspeyresia pactolana, Epinotia tedella,* and *Zeiraphera griseana.*[46,172] From Bavaria a number of insects are reported as novel pests. *C. abietis* and *Pristiphora abietina* occur in large numbers in spruce stands; *Bupalus piniaris* and *Diprion pini* in pine stands. None of these insects had been known to proliferate to this extent before the late 1970s. *Lymantria dispar* has also had mass outbreaks a number of times since 1976, although this pest had earlier been considered unimportant in these areas.[167]

In Austria, the sawfly *Pristiphora abietina* feeding on Norway spruce has become an increasing problem over the last 25 years and outbreaks are now seen at higher altitudes than before. Preliminary results from field studies in Hausruck indicate that air pollution, measured as SO_2 and NO_x, may play a major role in the outbreak pattern, while microclimatic differences could be ruled out as causes of differential outbreak between stands. Furthermore, a significant positive correlation could be shown between the content of SH-compounds in the needles in October and the number of wasps the following year.[12] In a similar study of the incidence of adelgids (*Sacciphantes* sp.) in a valley near Innsbruck, a correlation was established between the number of galls on Norway spruce and the general exposure to air pollution as determined from topographic parameters. In this case a significant correlation between the total sulfur content of the needles and the number of galls was shown.[165] Berge[11] reports a similar correlation between the sulfur content of pine needles and the degree of damage by the lepidopteran *Blastethia turionella.*

In Czechoslovakia, mass outbreaks of two gall midges (Diptera), *Harrisomyia vitrina* and *Drisina glutinosa* on sycamore maple were registered for the first time during the period 1980 to 1984, and the outbreaks were concentrated at high altitudes particularly subject to air pollution.[174] Preliminary studies of the pattern of attack by the balsam woolly adelgid, *Adelges picea,* on Fraser fir in the Appalacians also suggest that impact of polluted mist and rain may be responsible for the very high incidence of this insect at higher elevations.[75,76] Galling coccids (*Chermes viridis* and *Cephalodes abietis*) and bark-sucking aphids (*Dreyfusia nordmanniana*) are also reported to be a mounting problem in Poland,[171] while preliminary results from a survey of the coccid *Physochermes hemicryphus* on spruce in Germany showed no connection between attack and the general condition of the stands.[123] A reduction in the numbers of the aphid *Aphis frangulae* on alder buckthorn (*Frangula alnus*) in a pollution-affected area in Poland is attributed to a retardation of the phenological development of the trees, causing a disruption of the synchrony between hatching aphids and the bud burst of the host tree.[66] A comparative study of Dipteran communities in the industrial regions of Silesia and the agricultural region of Masuria in Poland revealed large differences in the structure and distribution of the communities. The number of families represented was reduced in the industrial regions polluted by SO_2 and heavy metals, and the trophic structure was modified towards an increase in the number and biomass of phytophages and a decrease in parasites and predators. Furthermore, a strong tendency for seasonal mass occurrence of phytophages could be observed in Silesia, indicating a disruption of the natural regulation of populations compared to Masuria.[38]

A comparison of the outbreak areas and patterns of *Epinotia tedella* over time as described by Cramer[37] and Führer[60,61] suggests that the distribution of this lepidopteran has changed from being most important in the Central European mountains toward more frequent attacks in lowland, industrial areas. The descriptions of the outbreaks also indicate a change from initiation in closed, dense stands of spruce toward more frequent initiation in the outer, exposed parts of the stands. Canadian surveys of the frequency and extent of outbreaks of the spruce budworm over the last 300 years have shown that the attacks in the 20th century occur more frequently, are more severe, and cover larger areas than earlier.[16]

III. STUDIES ALONG GRADIENTS

While the above-mentioned observations are often over large areas and time spans, their causes are still uncertain and will also be subject to underlying fluctuations in climate, changing forestry practice, etc. They do, however, suggest that atmospheric pollution may in some way be involved in altering herbivore-host relations. Studies along gradients of air pollution give further information on the underlying mechanisms and a firmer basis for claiming causal relationships.

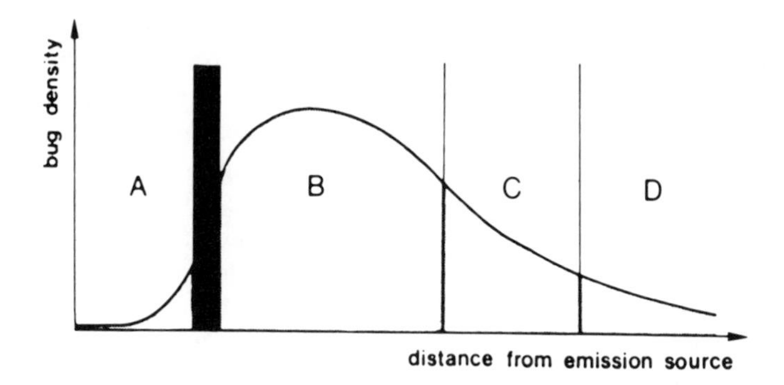

FIGURE 1. Schematized illustration of the occurrence of *Aradus cinnamomeus* along the gradient of pollutant concentration. A = lethal area with no or very few bugs, B = area of serious injury by *Aradus*, C = area of high population density without serious injury, and D = natural population density. (From Heliövaara, K. and Väisänen, R., *J. Appl. Entomol.*, 101, 469, 1986. With permission.)

In Finland, a series of surveys of the numbers of the pine bark bug *Aradus cinnamomeus* (Het., Aradidae) were performed along gradients from a fertilizer factory and copperworks emitting nitrogen oxides and sulfur dioxide, respectively, along with a number of heavy metals. The surveys showed that the density of the bugs was lowest right next to the factories, but rose quickly to a peak density at a distance of 1 to 2 km (Figure 1). At a greater distance, the population density slowly reverted to background levels. The densities in the heavily infested zone were up to 100 times the background level. Heavy attack by the pine bark bug in Finland is only seen in industrial areas.[85] Studies of the distribution of the pine resin gall moth *Petrova resinella* in the same area revealed a distinct gradient of increasing numbers of galls per tree with increasing proximity to the factory complex.[84] In connection with a mass outbreak of the sawfly *Pristiphora abietina* at the industrial town of Bodenfelde in Germany, it was shown that the number of wasps increased with increasing proximity to the SO_2-emitting industries.[196]

An extensive field survey of arthropod communities around two industries emitting HF and SO_2 in northwestern Russia showed a marked decline in the numbers of arachnids and myriapods along a gradient of increasing pollution. For the insects there was a trend towards species of relatively smaller body size with increasing pollution, but the numerical response was variable between the different insect groups. The general trend was an increase in sap-feeding species and a decrease in saprotrophic and predatory species with increasing air pollution. Both biomass and numbers of Coleoptera declined, while the numbers of Homoptera and Heteroptera increased with the pollution levels. Some species of lepidopterans were found to increase in the high-pollution zones, but the highest number of lepidopterans was found in the zone of intermediate pollution. The greatest numbers of the curculionid beetle *Hylobius abietis* were found in the zones of intermediate or low pollution. Within these zones the number was maximal in stands that showed marked deterioration but had not accumulated toxicants in the needles above a certain threshold. The numbers of this beetle peaked at a distance of 8 to 9 km from the aluminium plant and 5 to 6 km from the forest industry complex. The biomass of the individuals increased with distance from both industries.[108]

In a similar French survey in the Roumaire forest, the sample sites were characterized into pollution zones according to the epiphyte flora and the content of F and S in the needles of pine. The total number of insects was highest in the zones of intermediate pollution, while the numbers of aphids and the lepidopteran *Rhyacionia buoliana* were significantly higher in the zones of high pollution.[29] A later detailed study of the aphids in the same area revealed that *Cinara pini* L, *C. Pinea* Mordv., and *Protolachnus agilis* Kalt. were significantly more abundant in the zones of high pollution, whereas the opposite was true for *Scizolachnus tomentosus* Deg. and *Pineus pini* L.[191]

Templin[182] analyzed insect herbivory on *Pinus sylvestris* along gradients from a brown-coal-burning factory complex. By counting the trees attacked in successive zones of 5 km, it could be shown that the distance of visible impact of the complex was 20 km to the east and 30 km to the southeast. The percentage of trees attacked by the lepidopteran *Rhyacionia buoliana* from zone I-VI was 48.3, 32.7, 14.3, 12.6, 5.5, and 4.3%. There were also signs of increased attack by *Exoteleia dodecella* and *Petrova resinella* in the fume zone, while on the other hand the wood-boring insects *Myelophilus piniperda* and *Phaenops cyanea* were twice as numerous outside the smoke zone as close to the factories.

The aphid *Metapolophium dirhodum* on barley plants grown along a gradient of increasing SO_2 and NO_2 pollution from Ascot 35 km west of central London into central London itself showed a strong trend of decreased growth rates with increasing distance from the city.[133] Studies of the populations of the cercopid *Philaenus spumarius* similarly showed a gradient of decreasing densities downwind from a steelworks coke plant in northern England (ibid).

A number of other gradient studies have been performed along major roads. Przybylski[152] found a reduction in the total number of arthropods with increasing proximity to a motorway. The number of spiders, ladybirds, lepidopteran larvae, Elateridae, Hymenoptera, and Curculionidae increased with distance from the road, whereas the number of aphids and Heteroptera was greatest close to the motorway. The surveys were carried out in winter wheat, meadows, and orchards. A similar English survey showed an increase in both aphids and larvae of the gold tail moth *Euproctis simili* (Lep.) and the buff tip moth *Phalera bucephala* (Lep.) close to a motorway.[150] A series of Swiss studies have shown gradients in the numbers of *Aphis fabae* on *Viburnum opulus*[17] and *A. pomi* on *Crataegus monogyna*.[55] In both cases the number of aphids was highest close to a motorway. Gradients along motorways should, however, be interpreted with some caution, as the road influences microclimate and also works as a barrier to the mobility of both insects and other animals.[124] Furthermore, there is often a strong impact from deicing salt, which in itself can make the plants more susceptible to insect attack.[22,53]

The occurrence of melanic forms in industrial areas has long been known and seems to be connected to air-pollution-induced changes in the epiphytic cover on the tree trunks, as shown in studies of melanism in the lepidopterans *Biston betularia* and *Gonodontis bidentata* along a transect from Manchester to North Wales.[15] The changes involve a complex selection including differential predation on the color morphs. Whether there are differences in behavioral or physiological effectiveness of different color morphs is not known, but air pollution could be a powerful selective force on insect communities by giving an advantage to species exhibiting the capability of producing melanic forms or other fitness traits important in a polluted environment. It is evident from the field observations, as summarized in Table 1, that there is a great deal of variation in the response among different species of insects. Some species are enhanced by air pollution, others are inhibited, and still others seem to be favored by intermediate levels of air pollution. Each of these responses will be a product of many different factors. Führer[63] noticed that the great majority of novel pests enhanced by air pollution in forests are internal feeders upon which a direct impact from gases and dry deposited toxins may be very limited, while they at the same time may be very sensitive to favorable changes in the nutritional value of the host plant. Furthermore, species that tend to

Table 1
FIELD OBSERVATIONS OF CHANGES IN PHYTOPHAGOUS INSECT FAUNA IN CONNECTION WITH AIR POLLUTION

Insect	Host plant biotope	Pollution	Observation	Ref.
Heteropterans				
Heteropterans in general	Forest/meadows	Mixed industrial/roadside	+	108, 152
Aradus cinnamomeus	Pine	NO_x, SO_2	− + −	85, 168, 171, 172
Coptosoma scutellatum, Euryopicaris nitidus, Elasmucha grisea	Pine	HF, SO_2	+	108
Homopterans				
Homoptera in general	Forests/meadows	Mixed industrial/roadside	+	108, 152
Philaenus spumarius	Thistle	SO_2	+	133
Aphids in general	Forest/grassland	SO_2, NO_x, HF	+	90, 108, 150
Woolly aphids	Pine	SO_2, HF	−	11
Dreyfusia spp.	Fir	HF/mixed industrial	+	11, 148, 171
Sacchiphantes spp.	Spruce	HF/mixed industrial	+	149, 165, 184, 194
Cinara spp.	Spruce/pine	HF, SO_2	− + −	170, 171, 191
Adelges spp.	Fir/spruce	Mixed industrial	+	46, 76, 149
Protolachnus agilis	Pine	HF, SO_2	− + −	191
Protolachnus tomentosus	Pine	HF, SO_2	+	29
Schizolachnus tomentosus, Pineus pine	Pine	HF, SO_2	−	29, 191
Chermes viridis, Cnaphalodes abietini	Spruce	Mixed industrial	+	170, 171
Aspidiotus abietis	Spruce	Mixed industrial	+	170
Aphis frangulae	*Frangula* sp.	Mixed industrial	−	65
Aphis fabae	*Viburnum* sp.	Roadside	+	17
Aphis pomi	*Crataegus* sp.	Roadside	+	55
Hymenopterans				
Hymenopterans in general	Meadows/orchards	Roadside	−	152
Cephalcia abietis	Spruce	Mixed industrial	+	149, 171
Pristiphora abietina	Spruce	SO_2	+	12, 171, 172, 196
Pachynematus scutellatus	Spruce	Mixed industrial	+	171, 172
Acantolyda nemoralis	Pine	SO_2	−	169
Apis mellifera	?	SO_2	−	90
Lepidopterans				
Lepidopterans in general	Pine/aspen	HF, SO_2	− + −	108
	Meadows/grass	Roadside	−	152
Epinotia tedella	Spruce	Mixed industrial	+	37, 46, 172
Exoteleia dodecella, Rhyacionia buoliana	Pine	Mixed industrial	+	29, 149, 168, 172, 182
Zeiraphera griseana	Larch	Mixed industrial	+	171, 172
Laspeyresia pactolana	Spruce	Mixed industrial	+	172
Petrova resinella	Pine	Mixed industrial	+	84, 168, 171, 182

Table 1 (continued)
FIELD OBSERVATIONS OF CHANGES IN PHYTOPHAGOUS INSECT FAUNA IN CONNECTION WITH AIR POLLUTION

Insect	Host plant biotope	Pollution	Observation	Ref.
Blastethia turionella, Lymantria monacha	Pine	SO_2	+	11, 169
Phalera bucephala	Beech	Roadside	+	150
Euproctis simili	Hawthorn	Roadside	+	150
Dioryctria mutatella	Pine	Mixed industrial	+	172
Bark beetles				
Pityokteines sp.	Fir	HF/mixed	–	46, 148
Pityogenes chalcographus	Spruce	SO_2/HF	–	64, 171
Myelophilus piniperda	Pine	SO_2	–	182
Pissodes spp.	Conifers	Mixed industrial	+	148, 149, 171
Phaenops cyanea	Pine	SO_2/mixed industrial	+/–	19, 149, 171, 182
Ips typographus	Spruce	SO_2	+	19, 171
Ips amitinus	Pine	Mixed industrial	+	171
Dendroctonus sp.	Pine	Oxidants	+	136, 178
Hylurgops palliatus	Spruce	Mixed industrial	+	171
Acanthocinus aedilis	Pine	Mixed industrial	+	171
Diptera				
Thecodiplosis brachyntera	Pine/spruce	HF/SO_2	+	171, 172
Phytomyza ilicis	*Ilex aquifolium*	SO_2	–	11
Other groups and species				
Grasshoppers	Grass	Roadside	+	150
Coleoptera in general	Pine	HF, SO_2	–	108
Hylobius abietina (Col.)	Pine	HF, SO_2	+	108
Arachnids in general	Pine/meadows	HF, SO_2, NO_x	–	108, 152
Olygonychis ununguis (Arachnidae)	Pine/fir	Mixed industrial	+	171

Note: + = increases in populations,
– = decreases in populations, and
– + – = a domeshaped response with maximal densities at intermediate levels of air pollutants.

prefer dying tissue (secondary pests) would be expected to benefit from air pollution, whereas insects belonging to the primary pests may react negatively to severe stress on the host plant.

Führer[64] has provisionally grouped a number of insects according to their observed response in polluted environments. The following is based on his work but supplemented with further examples.

Type A: — Species that are definitely favored by strong air pollution and thus occur in frequent or chronic mass outbreak in such areas. Examples are the lepidopterans *Exoteleia dodecella, Rhyacionia buoliana, Petrova resinella,* bark beetles of the *Pissodes* family, the hymenopteran *Pristiphora abietina,* and most aphids.

Type B: — Species that are more or less absent from the center of the pollution zone but often found in elevated numbers in zones of Intermediate pollution. Examples of this type may be the aphids *Cinara* sp. and *Protolachnus agilis;* the bark bug *Aradus* *cinnamomeus;* the curculionid beetle *Hylobius abietis;* and perhaps the bark beetles *Phaenops cyanea, Ips sexdentatus,* and *I. typographus.*

Type C: — Species that are common in unpolluted environments but virtually absent from polluted areas. Examples are the bark beetle *Pityokteines curvidens* and a number of coleopterans and hymenopterans, as exemplified by *Acantolyda nemoralis* in the case of SO_2 pollution.

These groupings are, however, based on few field studies and much more work is needed before any definite conclusions can be drawn on the response of different insects to air pollution. This is accentuated by the fact that the pollution involved in the different studies varies a great deal and the response of the insects may be very much determined by the specific combination of pollutants, as well as by complex differential effects on predators and parasites.

IV. EXPERIMENTAL EVIDENCE OF INTERACTIONS

While observations in the field are often difficult to relate directly to air pollution because of the presence of a number of other variables, a number of controlled experiments have demonstrated the capacity of air pollution to change plant-insect interactions (Table 2). The experimental work covers only a few groups of insects, most of which are economically important pests, and this bias towards a few species presents a danger of drawing general conclusions that may not hold for other groups of insects. The bulk of work has been done on aphids which have the advantage of being easy to culture and handle and at the same time have very fast rates of growth and reproduction, which makes short-term experiments feasible.

A. APHIDS

Dohmen et al.[45] showed a significant increase in the mean relative growth rate (MRGR) (calculated from the intial and final weight of the aphids) of *Aphis fabae* feeding on *Vicia fabae,* which had been prefumigated for 7 d with either 148 ppb SO_2 or 204 ppb NO_2. Fumigation of aphids living on artificial diet sachets showed no direct effect of the gases on the MRGRs of the aphids. Since the tested concentrations are somewhat above mean values recorded in ambient air in Britain, the system was additionally fumigated with either ambient London air (13 to 54 ppb SO_2 and 10 to 50 ppb NO_2) or charcoal-filtered air. The results showed that the aphids in the ambient air chamber had a significantly higher growth rate than aphids from the filtered chamber.[45] A later experiment showed that the aphid *Macrosiphon rosae* grown on *Rosa* sp. in chambers with ambient Munich air had 20% higher MRGRs than in a filtered chamber. The average mean half-hour values in the ambient air were 9.25 ppb SO_2, 20 ppb NO_2 and 12 ppb O_3.[44]

Similar results have been obtained by Swiss researchers with *A. fabae* on *Viburnum opulus* and *Phaseolus vulgaris,* and with *A. pomi* on *Crataegus* species at the verge and central reservation of a motorway. When *Crataegus monogyna* and *V. opulus* plants were enclosed in chambers with filtered air, the number of aphids close to the motorway was reduced compared to a control chamber, indicating that the enhanced aphid growth is partly caused by the polluted air and not only by effects of temperature, deicing salt, etc. The NO_2 concentrations in the *Viburnum* experiment were 416 ppb and in the *Crataegus* experiment between 255 and 612 ppb.[17,18,23] In another experiment, the native aphids on *C. monogyna* were killed by spraying them with Veralin-D and the plants were reinfested with five aphids per shoot. The polulation growth rates on sites in the

Table 2
EXPERIMENTAL EVIDENCE OF INTERACTIONS BETWEEN AIR POLLUTION AND INSECT HERBIVORES

Insect	Host plant	Treatment	Reaction	Ref.
Aphids				
Aphids fabae	*Vicia faba*	SO_2, NO_2, ambient air	Increased growth rate	45, 133
	Viburnum opulus	Ambient air	Increased population	17
Aphis pomi	*Crataegus* sp.	Ambient air	Increased population	23
Macrosiphon rosea	*Rosa* sp.	Ambient air	Increased growth rate	44
Acyrthosiphon pisum	*Pisum sativum*	SO_2	Increased growth rate	192
	Vicia faba	SO_2/NO_2	Decreased growth rate	133
Sitobion avenae	Wheat	SO_2	Increased growth rate	133
Rhopalosiphum padi	Wheat/barley	SO_2/NO_2	Increased growth rate	133
Metapolophium dirhodum	Wheat/barley	SO_2/NO_2	Increase growth rate, increased density	133
	Barley	Gradient of ambient air	Increasing growth rates	133
Macrosiphum albifrons	*Lupinus* sp.	SO_2	Increased growth rate	133
Myzus persicae	*Brassica oleracea*	SO_2	Increased growth rate	133
Brevicoryne brassicae	*Brassica oleracea*	SO_2	Increased growth rate	133
Phyllaphis fagi	Beech	Ambient air	Increased population	53
Elatobium abietinum	*Picea sitchensis*	SO_2	Increased growth rate	133
Other insect groups				
Epilachna varivestis (Mexican bean beetle)	Soybean	SO_2	Increased growth, increased fecundity, decreased mortality, preference for treated plants; increased population	98, 100
	Soybean	O_3	Increased growth, preference for treated foliage	28 48

Table 2 (continued)
EXPERIMENTAL EVIDENCE OF INTERACTIONS BETWEEN AIR POLLUTION AND INSECT HERBIVORES

Insect	Host plant	Treatment	Reaction	Ref.
	Pinto bean	SO_2	Preference for treated plants, but no effects on growth and reproduction	97
Lymantria dispar	*Quercus alba*	O_3	Preference for high-ozone foliage, reduced consumption at intermediate levels	105
Keiferia lycopersicella	Tomato	O_3	Increased survival, faster development	189
Grasshoppers	Mixed grass	SO_2	Decreased density	131

central reserve, at the verge of the motorway, and at a control site were similar to those of the natural populations, indicating that the difference in aphid numbers was not caused by a higher initial number of eggs close to the motorway.[21]

The aphids *A. fabae* on broad beans (*Vicia faba*) and *Sitobion avena* on wheat (*Triticum aestivum*) showed increases in MRGR when feeding on the plants during fumigation with 100 ppb SO_2.[133] Similar results were obtained for *Acyrthosiphon pisum* on *Pisum sativum, Macrosiphum albifrons* on *Lupinus* spp., *Myzus persicae* on *Brassica oleracea,* and *Rhophalosiphum padi* on wheat and barley. Fumigations with a combination of 100 ppb SO_2 and 100 ppb NO_2 resulted in increased growth of *Aphis fabae* on *V. faba* and *S. avenae* and *Metapolophium dirhodum* on wheat and barley. In all these studies the aphids were feeding on prefumigated plants and not subjected to the gases themselves (ibid.).

Warrington[192] has investigated that response of the pea aphid *Acyrthosiphon pisum* on *P. sativum* to various concentrations of SO_2 in environmental chambers. Aphids were allowed to feed for 4 d during fumigations and MRGRs were calculated. The MRGR increased with increasing SO_2 levels to a peak of 11% above the controls between 90 and 110 ppb. At higher concentrations the increase in MRGRs leveled off and at 200 ppb SO_2 they were below those of the controls (Figure 2). The MRGRs were significantly higher than the controls in the 80 to 130 ppb range. In the range from 0 to 105 ppb SO_2 there was a significant linear correlation between the SO_2 level and both absolute and relative increases in MRGR, and the curve suggested that there may not be any lower threshold level for the effect of SO_2 on aphid growth. In a later experiment, the aphids were introduced after the termination of the fumigation and left to feed for 4 d on the prefumigated plants. In this experiment the enhanced growth continued above 110 ppb, indicating that the decline in MRGRs above 110 ppb in the previous experiment was due to some direct effect of the gas on the aphids rather than mediated through

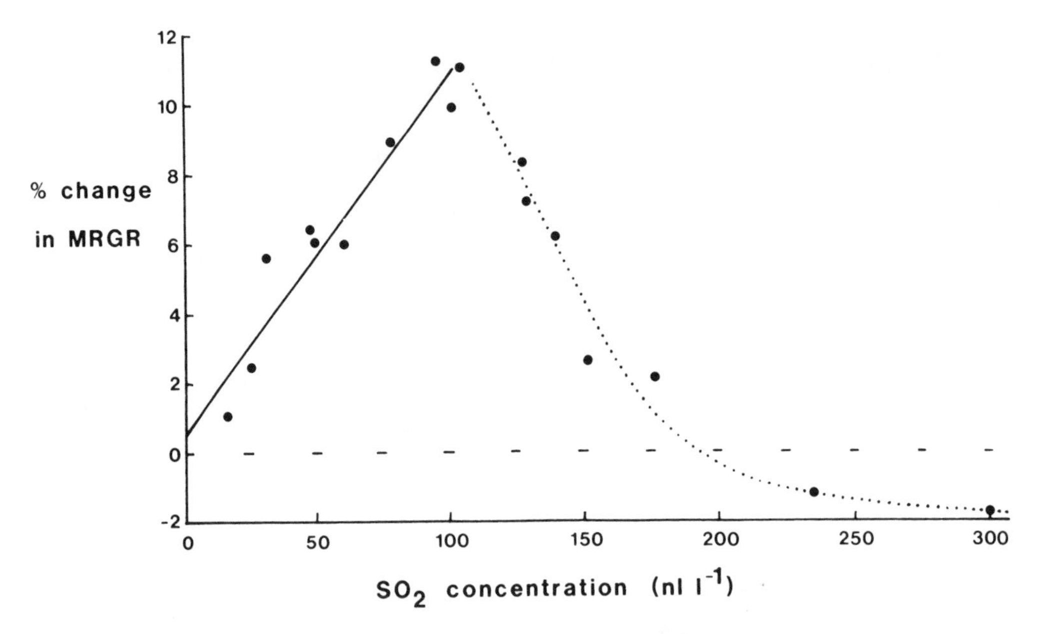

FIGURE 2. Relationship between percentage change, relative to controls, in MRGR of pea aphids feeding on pea plants and SO_2 concentration. Plants and aphids were fumigated for 4 d. Solid line: linear regression relationship for range 0 to 105 nl/l SO_2 ($r = 0.958$, $p < 0.001$); dotted curve fitted by eye. (From Warrington, S., *Environ. Pollut. Ser. A,* 43, 155, 1987. With permission.)

the plants.[193a] McNeill et al.[133] found a similar dome-shaped response curve for *Aphis fabae* feeding on *V. faba* after a 7-d prefumigation with SO_2. The MRGRs rose from 0.51 μg/μg/d at 0 ppb SO_2 to a peak of 0.54 to 0.56 in the range of 80 to 150 ppb SO_2 and decreasing to 0.50 on plants prefumigated at 250 ppb SO_2. In this case direct effects of the gas on the aphids could be excluded.

The two important cereal aphids *S. avena* and *M. dirrhodum* were monitored from 1983 to 1986 in an open fumigation system with winter wheat (1983 to 1984) and winter barley (1984 to 1985) as host plants. In all years there were significantly higher populations of aphids on the higher SO_2 treatments compared to the controls in July prior to the collapse of the populations upon ripening of the cereals. The population growth of both species increased with increasing SO_2 levels (9, 21, 32, and 43 ppb).[133] Similar increases in reproduction have been demonstrated by Warrington et al.[193] for *Acyrthosiphon pisum* growing on *P. sativum* during fumigations with 50 and 80 ppb SO_2. The production of nymphs increased by 19% in the fumigated chambers relative to controls, resulting in a 4.6% increase in the intrinsic rate of population increase (r_m).

There have been few studies with aphids on forest trees. In one experiment the green spruce aphid *Elatobium abietinum* was allowed to feed on potted saplings of *Picea sitchensis* prefumigated with 100 ppb SO_2 for various periods of time. The MRGR of the aphids showed a domed response rising from a mean of 0.258 μg/μg/d in clean air to a peak of 0.308 after 3 h exposure of the plants and falling back to 0.29 at expsoures of 16 to 48 h duration.[133] Exposure of cottonwood (*Populus deltoides*) to acute doses of ozone (200 ppb) for 5 h did not significantly change the susceptibility of the trees to the aphid *Chaitophorus populicola,* but the authors suggest that a different dose may cause effects.[35] Flückiger and Braun[53] found an increase in populations of *Phyllaphis fagi* on beech seedlings exposed to ambient, rural air compared to filtered air in open-top chambers. Details are not given about the constituents of the ambient air, but an impact of ozone is most probable in this region of Switzerland.

There is thus strong evidence that air pollution at ambient levels may enhance the growth of a wide range of aphids, and this is in good accordance with a large number of field surveys where sucking insects have been seen to proliferate in areas affected by air pollution. The only recorded example of inhibition of aphid growth on fumigated plants was a reduced growth of *A. pisum* on *V. faba* exposed to either 100 ppb SO_2 and 100 ppb NO_2 or ambient air compared to the performance on clean air controls.[133]

B. OTHER INSECTS

SO_2 has been shown to have an impact on a number of other phytophagous insect species. McNary et al.[131] reported a significant reduction in the number of grasshoppers on mixed-grass prairie plots over a 3-year period of fumigation with SO_2 in an open fumigation system. The response was shown at two different sites and seemed to be concentration-dependent with mean densities of grasshoppers of 2.9, 2.4, 1.9, and 1.8/m^2 for plots of 0.52 ha receiving $<$10, 28, 52, and 87 ppb SO_2, respectively (monthly mean values). The experimental design did not permit any distinction between direct effects on reproduction and survival and indirect effects of changing immigration/emigration ratios.[131] Subsequent feeding trials did, however, show a preference for control leaves over SO_2-fumigated leaves.[115]

A number of experiments have shown that SO_2 fumigation of soybeans stimulates growth and reproduction of the Mexican bean beetle (*Epilachna varivestris*). In one experiment, the beetles were fed excised leaves from soybean plants fumigated for 7 d with 200 ppb SO_2. Beetles feeding on fumigated leaves had significantly shorter development times and reached a greater weight. Furthermore, females feeding on fumigated leaves laid up to twice as many eggs per female as did beetles feeding on control leaves. Adult females had a highly significant preference for fumigated leaves ($p < 0.001$).[98] The increased growth on fumigated plants was primarily caused by an increase in consumption, while changes in the nutritional value of the leaves apparently was of little importance.[94] In a later experiment, the beetles were introduced to prefumigated soybean plants grown in cages in the field and left to feed during continued fumigations with 136 ppb for 191 h during a 68-d period, giving a seasonal average of 16 ppb SO_2. The number of beetles in the fumigated field cages was 1.5 times that of the control cages. This was largely due to a significant increase in female fecundity in the fumigated cages, while survival was the same in the two treatments. The larvae also reached a greater weight on the SO_2-fumigated plants.[100] There may have been an interaction with ozone in this experiment, since SO_2 was added to ambient air with an average ozone content of approximately 35 ppb. The concentrations used in these experiments are somewhat higher than usual ambient levels, but are not unrealistic around point sources or during pollution episodes (inversions, etc.). When soybean leaves were fumigated for 24 h at concentrations of 25, 50, 100, 300, 500, and 700 ppb SO_2 and fed to Mexican bean beetles, all treatments above 25 ppb gave rise to increased growth of the larvae. The response peaked at 300 ppb and the effect leveled off at higher concentrations. The same response pattern was found with plants that had been allowed to recover for 24 h before being fed to the beetles, although the weight increases of the beetles were lower in this instance.[101] An earlier experiment in which the beetle *E. varivestis* was fed leaves of pinto bean prefumigated for 7 d with 150 ppb SO_2 showed a significant preference for fumigated leaves, but there were no effects on growth, fecundity, or survival.[97] A chamber experiment using the two-spotted spider mite *Tetranychus urticae* Koch. on soybean showed no effects of SO_2 fumigation.[99]

Relatively few experiments have been conducted on the effect of ozone fumigation on insect herbivores. Endress and Post[48] have demonstrated that Mexican bean beetles in the laboratory prefer to eat soybean leaves prefumigated with ozone-enriched ambient air compared with leaves grown in ambient air. The ozone levels used for prefumigation were 25, 50, 78, and 114 ppb on each of 16 occasions of 7 h each when the plant age was 17 to 53 d. Paired tests of feeding preference indicated a dosage-dependent response. The difference in feeding preference between the 50 and 114 ppb prefumigated plants was present even after 16 and 23 d following the termination of fumigations.

An analogous experiment in which the larvae of the gypsy moth *Lymantria dispar* (Lep.) were fed oak leaves prefumigated with ozone gave a more complex response. The oak seedlings were fumigated in 11 exposures of 11 h each with ambient and ozone-enriched air giving concentrations of around 35, 90, and 135 ppb during exposure. In the preference test the larvae chose the leaves prefumigated at 135 ppb over the controls and the 35 ppb, while they preferred 35 ppb and the controls over the 90 ppb leaves. This indicates that moderate concentrations of ozone may lead to biochemical changes in the oak leaves that make them less palatable to the larvae, while high concentrations improve palatability.[105] Trumble et al.[189] found significant increases in survival and decreases in development time of tomato pinworm *Keiferia lycopersicella* (Lep.) on ozone-treated tomato plants. The plants were treated two or four times at 3-d intervals with 280 ppb ozone for 3 h. The survival was higher on plants fumigated only prior to infestation than on plants fumigated while the larvae were feeding as well. Whether this was due to a direct toxicity of the ozone to the larvae was not ascertained, but the negative impact of the gas was reduced when the larvae entered their mining stage.

Chappelka et al.[28] looked at the response of Mexican bean beetle to soybean fumigated with O_3 in open-top field chambers for 3 weeks. The 7-h average O_3-concentrations were 21 to 24, 34 to 40, 71 to 74, and 94 to 99 ppb during three different experiments. After fumigation the plants were transferred to a coal-filtered chamber and the beetles were introduced to a mixture of plants from the four fumigations. After 6 to 8 d a positive linear correlation could be shown between the degree of defoliation and the previously experienced ozone concentrations. Defoliation was an estimated 20.5% in plants from the high-ozone treatment compared to 10% in the controls. Development time decreased and larval weight increased when beetles were fed O_3-treated foliage in the laboratory. The fastest development was obtained on leaves treated with intermediate ozone levels. This was probably due to necrotic injuries on the high-ozone plants.

C. SYNERGISTIC EFFECTS

Interpretations of results involving necrotic injuries and insect feeding on plants in the field and in experiments are complicated by the fact that in some cases there seem to be synergistic interactions whereby plant damage by insects and polluting gases is greater than that expected by either alone. Rosen and Runeckles[159] showed, in an experiment with the whitefly (*Trialeurodes vaporarium*) on bean plants, that infested plants developed more necrotic flecks than uninfested plants when fumigated with 200 ppb ozone for 6 h/d for 6 d. The effect is probably caused by an increased access of ozone to the mesophyll of leaves that have been injured by the insects. Berge[11] has reported similar observations. *Picea omorika* trees that had been sprayed with an insecticide did not develop any necrotic injuries during a pollution episode with SO_2 and HF, while the unsprayed trees, infested with the mite *Tetranychus ununguis,* were severely injured by the high levels of pollutants. Likewise, the holly (*Ilex aquifolia*) damaged by the dipteran leaf miner *Phytomyza ilicis* was more injured by the pollutants than the uninfested plants. Interestingly, the leaf miners seemed to prefer plants without necrotic injuries at the time of colonization in the spring. In the case of insecticide treatment, however, there may have been other factors involved; for instance, antioxidant properties of the insecticide providing the plant with a protection against injury.

D. ACID RAIN

Although it seems likely that acid rain could play a role in the observed changes in insect populations in the field, very few experiments involving acid precipitation or soil acidity have been performed. In an ongoing Finnish experiment, where pine and birch trees are being treated in the field with artificial acid rain of pH 3 and pH 4, the number of lepidopterans, sawflies, and aphids on the trees are being recorded. No dramatic changes have been observed, during the 2.5 years of the experiment, but the insects seem in general to be favored by the acid treatment, significantly so in a few cases.[80a] Neuvonen and Lindgren[138] have shown that acid treatment of foliage and soil at pH 3.5 makes birch trees a better substrate for the aphid *Euceraphis betulae.* The effect was most pronounced during periods in which the precipitation was below average, which indicates an interaction between drought and acid stress. An experiment in which *Rumex hydrolapathum* was grown in hydroculture at different pH levels showed that growth and reproduction of the butterfly *Lycaena dispar batava* was generally reduced at pH levels below 5.5, and the reduction corresponded with a reduction of the content of N, P, and K in the foilage.[14]

The mechanisms underlying the changes observed both in the field and in experiments can be grouped as follows:

1. Direct effects of the gases, acid rain, or deposited toxins on the insects
2. Pollution-induced changes in the balance between herbivores and their predators and parasitoids
3. Changes in the suitability of the host plants

These mechanisms are reviewed separately in the following sections.

V. DIRECT EFFECTS OF POLLUTANTS ON INSECTS

Part of the differential response between insect species may be caused by variable sensitivity to the direct effects of toxicants. Only a few insect species, however, have been investigated for tolerance to air pollutants and few of these are herbivores. This is understandable since the insects must be reared on artificial diets in order to isolate the direct effects of the pollutants from the indirect effects due to changes in the host plant. Experimental work is thus extremely difficult on aphids and other insect groups for which satisfactory artificial diets have rarely been developed.

Only occassionally have atmospheric pollutants been found to have a direct positive effect on insects. Beard[8] and Levy et al.[120] found that low levels of ozone greatly stimulated oviposition of houseflies (*Musca domestica* L.) which could lead to a significant increase in the adult population. Feir and Hale[50] reported that when the large milkweed bug *Oncopeltus fasciatus* was exposed to 10 ppm SO_2 for 2 h/d, egg production and growth rates were stimulated and longevity of adults was increased. However, this last result contrasted with earlier experiments, where *O. fasciatus* was exposed to combinations of SO_2, NO, NO_2, CO, O_3, and hydrocarbons at levels simulating those in American cities. Over a range of concentrations, the bugs showed reduced growth, smaller size at maturity, and reduction in egg production, although this was less true of bugs reared in polluted air than those transferred from clean to polluted air.[51] This is in line with the findings for fruitflies (*Drosophila melanogaster*) treated with 0.7 ppm SO_2 for 11 d,[68] while male sweat bees (*Lasioglossum zephrum*)[69] and honeybees[2,89] both showed reduced flight activity when subjected to SO_2. Populations of honeybees were reduced in the vicinity of a coal-fired power plant emitting SO_2.[89] The oribatid mite

Humerobatus rostrolamellatus (Grandjean) was also found to be very sensitive to SO_2.[113] These observations show that SO_2 in high concentrations and even sometimes at ambient levels can be harmful to various insects. On the other hand, some species seem to be extremely tolerant of air pollutants. Petters and Mettus[147] fumigated females of the ectoparasitic hymenopteran *Bracon hebetor* with 3000 ppb SO_2 for 3 h and 5 h. Even at these very high concentrations, no differences in survival or egg production could be shown between treated insects and controls. Lebrun et al.[113] found no changes in the mortality of the psocid *Hyperetes guestfalicus* and the coleopterans *Lema cyanella* and *Plagiodera versicolor* when exposed to 1500 ppb SO_2 for several days. McNeill et al.[133] found no effects of 100 ppb SO_2 on the growth of the aphids *Sitobion avenae* and *Aphis fabae* reared on an artificial diet. Exposure of the cockroaches *Periplaneta americana* and *Nauphoeta cinerea* and the ant *Solenopsis invicta* to 300 ppb ozone for several weeks revealed no effects on mortality.[121]

The effects of nitrogen oxides on insects have only been tested in a few experiments. Dohmen et al.[45] found that *A. fabae* fumigated with 204 ppb NO_2 while feeding on artificial diet sachets had MRGRs similar to controls living in filtered air. Feir and Hale[50] found that NO_2 in high levels (5 to 10 ppm) increased the mortality of the milkweed bug *O. fasciatus,* but the surviving bugs had significantly higher growth rates than the controls. The average number of eggs per female was reduced by the treatment.

The general picture that emerges from the scattered experiments on direct gas effects on arthropods is that there seems to be a highly variable sensitivity to pollutants between species. This suggests that direct effects of gaseous pollutants could be a driving force in changes in community structure. There is great need for more experiments to elucidate the effects of combinations of gases and to separate direct effects of the gases from effects of ingestion of dry deposited toxicants. In addition, differences in other conditions during experiments such as humidity and temperature could be important. Bioaccumulation of toxins as implied by the results of Dewey[43] could lead to changes in the balance between herbivores and predators, but there is a total lack of experimental evidence for this mechanism.

VI. CHANGES IN PARASITISM AND PREDATION

A number of field studies imply that changes in parasitism and predation may play a role in the release of herbivorous insects from innocuous levels in polluted areas. Templin[182] showed a reduction in the number of parasitized larvae of *Rhyacionia buoliana* with increasing proximity to a coal-fired power plant, and the trend could be seen during 2 successive years. Polish studies have shown similar reductions in parasitism in *Exoteleia dodecella* (Lep.) in industrial areas,[169] and in Brazil the proportion of the lepidopteran *Papilio scamander scamander* parasitized by the hymenopteran *Spilochalchis* sp. was lower in the city center of Porto Alegre compared to suburban areas.[160] Wentzel and Ohnesorge[196] likewise observed that the increased attack of the sawfly *Pristiphora abietina* near an SO_2 source was partly caused by a lower mortality of the wasps in the pupal stage, and that this could be related to a decrease in the percentage of parasitized wasp pupae in the pollution zone. Hillmann and Benton[90] found a reduction in the number of parasitic wasps near an SO_2-emitting power station in Pennsylvania and speculate that this may in part explain a concurrent increase in the number of aphids. Villemant,[191] however, found no significant differences in the proportion of parasitized aphids on spruce in areas with different degrees of air pollution, while the aphid numbers were highest in the zones of high pollution.

Finnish gradient studies have shown a significant negative correlation between the proportion of eggs of the pine bark bug *Aradus cinnamomeus* parasitized by the hymenopteran *Telonomus aradi* and the distance from an iron factory. The decreased level of parasitism close to the factory corresponded with a significant increase in the number of bugs.[87] A later study of parasitism on *Petrova resinella* galls around a factory complex, however, did not show any correlations between the proportion of parasitized galls and the degree of pollution.[86] Austrian observations suggest that the degree of parasitism on the sawfly *Cephalcia abietis* by the endoparasitic nematode *Steinermema kraussei* may be altered by acidification of the soil. A transect outward from tree trunks showed an increase in the number of parasitized pupae with increasing distance from the trunk. The nematodes are known to be important in the population dynamics of *C. abietis*.[52] The acidity is, however, not the only factor that may influence the nematodes in the zone of stem flow and experimental evidence is needed for conclusions to be drawn.

Experimental evidence of effects of air pollution on parasitism has been shown in two studies. In an open field fumigation of winter wheat and winter barley with SO_2 there was clear evidence of a reduction in the proportion of parasitized aphids in the plots receiving SO_2 at levels of 9 to 43 ppb. Whether this was caused by a direct effect of the gas on the behavior of the parasites or simply an effect of higher rates of increase in the aphid populations is not known.[133] Similar SO_2 fumigations of prairie grassland likewise revealed a reduction in the number of parasite and parasitoid species at SO_2 concentrations of 52 and 87 ppb monthly mean.[115]

The above examples show that changes in parasitism may be involved in changes in herbivore populations in polluted areas, but the evidence is still too fragmentary to allow for any general conclusions. This is especially true because none of the surveys has recorded the density of the adult parasitoids, but only made indirect counts via their activity. It is possible that some parasitoids (e.g., hymenopterans) suffer from direct effects of the pollutants, perhaps by way of reduced flight activity as shown for honeybees and sweat bees.[69,90] There is, however, a need for more detailed studies of these processes.

As can be seen from the field studies mentioned earlier, changes in the number of predators have been seen several times in polluted areas, but no detailed studies of predator-prey systems have been performed. Clausen[32] found a negative correlation between the number of taxa of spiders on elm trunks and the degree of SO_2 pollution, with the predatory species being most affected. Katayev et al.[108] found a general reduction in the numbers of Arachnida, Myriapoda, Staphylinidae, and Carabidae in a polluted area. Furthermore, a reduction in size and number of anthills (*Formica rufa*) was observed along a gradient of increasing air pollution. Studies along a gradient of industrial pollution (mainly SO_2) in Poland have likewise revealed reductions in the size of ant colonies (*Myrmica* sp. and *Lasius niger*) and a reduction in the mean biomass of the workers.[146] The number of *Formica* spp. is also decreasing in pollution-threatened forests of Czechoslovakia, possibly due to accumulation of heavy metals received through honeydew.[179] Comparative studies of ladybird (Coccinellidae) communities in Poland have shown that this important group of predators on aphids is reduced in numbers and species diversity in the industrial district of Silesia compared to similar biotopes in the Masurian lakeland.[65] The general trend of reduction in the number of beetles and arachnids caught in pitfall traps in polluted areas could also imply a general decrease in predation on pupae and prepupae in the soil as well, but the experimental evidence is lacking. Braun and Flückiger[21] could, however, show a difference in predation on aphids between the central reservation of a motorway and a control site. When *Crataegus* bushes were enclosed in open and closed cages at both sites, it was found that the number of aphids in the open cages at the control site was only 1.8% of that in the closed cages, whereas at the central reservation the population in the open cages averaged 32% of that in the closed cages.

A factor that may be of great importance to both parasitism and predation is the observed decrease in development time of a number of species of insects on fumigated plants, which will tend to reduce the probability of predation and parasitism by reducing the time in which the insects are vulnerable.[189] The observations on both parasitism and predation suggest that increases in herbivore populations in polluted areas may be governed by a dual mechanism involving release from innocuous levels by changes in predation/parasitism and increased growth and reproduction caused by changes in the nutritional value of the host plants.

VII. CHANGES IN THE HOST PLANT

Air pollution has been shown to change the biochemical composition of many plants and hence the nutritional value of the plants to herbivores. Studies of the nutritional requirements of insects have been performed on intact plants or on controlled diets where only one or a few components are varied. Studies on whole plants are difficult to interpret, because nutritional value is a matter of balance between components and registered changes in one compound may often be accompanied by nonregistered change in another component of the diet. Artificial diets overcome this problem, but often lack the wide spectrum of secondary metabolites that may be important for insect feeding and nutrient uptake.

The picture emerging from the many studies is that there is very little variation in the qualitative demands for food between different insect species and that virtually all plants contain the essential nutrients. Differences in insect success on different plants must therefore be sought in the balance of the nutrients or in the interference of various secondary metabolites.[9,39,59,91] Despite the importance of balances between nutrients, it is possible to point out some nutritional factors as more important than others. A number of these factors are considered in the following paragraphs and known effects by air pollution will be outlined.

A. CARBOHYDRATES

Carbohydrates represent the most important source of energy for herbivorous insects, especially during the later larval stages.[141,190] Some carbohydrates are much more important than others: pentose compounds can generally not be utilized by insects[39] and may even inhibit growth;[59] hexose sugars are the most important source of energy for the larvae of the sawfly *Gilpinia hercynia,* but the uptake depends on the specific hexose (fructose is the only hexose utilized 100% from spruce needles); sucrose is utilized 100% and its content in the needles correlates with the cocoon weight of the sawflies.[161-163] Other studies have shown that sucrose causes an increase in the food uptake, for example in larvae of the spruce budworm *Choristoneura fumiferana*[129] and Mexican bean beetle.[3] Finally, not all insects are able to digest starch and it is a general notion that carbohydrates can only be taken up by the insect gut as monosaccharides.[59]

One of the primary effects of air pollution seems to be changes in the carbohydrate metabolism which can be observed even at rather low levels of pollutants. Barnes[7] found increases in both total soluble carbohydrates and reducing sugars after long-term fumigations of five species of pine seedlings with 50 ppb ozone. Tingey et al.[187] found similar increases in soluble sugars as well as starch in the above-ground parts of ponderosa pine seedlings fumigated with 100 ppb ozone for 20 weeks, while starch and sugars in the roots decreased. In a previous long-term exposure of pine seedlings to low levels of ozone, the amount of soluble sugars in the needles decreased at first, but later showed a significant increase.[186] Short-term exposure of young pine trees with high concentrations of SO_2 resulted in an increase in the amount of reducing sugars and a decrease in the amount of nonreducing sugars in the needles. The change is hypothesized to be caused by a combination of increased breakdown of starch reserves and a reduction in photosynthesis.[126] Similar increase in reducing sugars have been found in a gradient study of hawthorn along a motorway. The increase correlated with increasing vicinity to the motorway and an increase in aphids.[55] Grill et al.[72] found increased levels of free glucose in SO_2 damaged needles of *Picea abies* collected in the field compared to healthy needles.

Koziol[111] has reviewed a large number of experiments, and from this review it seems that increase in soluble sugars is a general response to both SO_2 and O_3, while the content of starch usually decreases in long-term exposures. Both increases and decreases in starch and soluble sugars have been observed in response to fumigations with HF. Changes of this sort may explain some of the reports of increased insect herbivory on pollution-stressed plants.

B. NITROGEN AND AMINO ACIDS

Experiments using artificial diets have in many cases shown a clear correlation between insect growth and the N content of the food.[26,91] Field studies and laboratory work using leaves and needles have shown similar correlations, although the results are more ambiguous. More-detailed discussions of the significance of nitrogen to herbivores have been published.[128,132,197]

Scriber[166] has summarized a large number of experiments and found that positive correlations between insect growth and N content have been found in 160 studies, while 44 studies have shown a negative correlation. A lack of correlation may be due to the fact that not all N compounds can be utilized by insects and it is the available nitrogen which is important.[198] The levels of soluble proteins and amino acids may be much more important than total nitrogen, particularly to sucking insects.[132,143] Also, the proportions of different amino acids have been correlated with performance of several homopteran species.[151]

Not many studies have focused on total nitrogen in relation to air pollution. Swiss studies found significant increases in total nitrogen in *Crataegus* spp. and both total N and dissolved amino acids in *Viburnum opulus* and *Phaseolus vulgaris* exposed to ambient air at a motorway compared to coal-filtered controls,[18,23] while Tingey et al.[187] found that 100 ppb ozone for 20 weeks did not affect the total nitrogen content of needles of ponderosa pine. Decreases in total nitrogen have been seen in a number of studies involving SO_2 fumigations.[114]

Changes in the amino acid content caused by air pollution have been shown in numerous studies. In connection with the already-mentioned Swiss observations of increased aphid attack along motorways, it was shown that the amounts of soluble amino acids of hawthorn and birch increased with increasing proximity to the motorway.[55,56] Later measurements of the protein content of birch and *Cornus sanguinea* showed an increase in protein close to the motorway in spring but a decrease compared to controls later in the season.[20] Equivalent English studies showed a slight increase in soluble amino acids and considerable increases in total protein content of beech leaves close to a motorway.[150] Trumble et al.[189] found that ozone-fumigated tomato plants enhancing the growth of tomato pinworm had increased levels of soluble protein and amino acids and that the composition of the amino acid pool was altered. Alanine, glycine, serine, and valine increased, while the levels of glutamine, methionine, proline, and tyrosine were reduced in the fumigated plants.

Ozone fumigations of bean plants have been shown to cause marked increases in both dissolved amino acids and the total protein content of the leaves.[186] The concentrations of ozone used, however, were much higher than average ambient concentrations (250 and 500 ppb). SO_2 fumigations of young pine trees have likewise led to increases in dissolved amino acids and a change in the composition of the amino acid pool. The changes in the amino acid metabolism occurred before any visible injuries could be observed. The authors suggest that the rise in amino acid levels may have been caused by hydrolysis of proteins.[126] Field observations in Germany imply that the same pattern may be working in Norway spruce. The content of dissolved amino acids increased considerably along a gradient of increasing exposure to SO_2, while the total protein content was reduced. The rise in amino acids could not be explained by protein hydrolysis alone, and it is suggested that it may be caused by an increased input of ammonium along with the SO_2.[200]

It thus seems that an increase in levels of soluble amino acids is a general effect of air pollution while the protein content may either increase or decrease. Both changes in total protein and in the composition of the amino acid pool seem to be important factors for insect growth and reproduction. It is remarkable that an increase in dissolved amino acids is also a common reaction to drought stress and believed to be a factor predisposing plants to insect attack.[129]

Apart from the total content of proteins and amino acids, some amino acids and peptides seem to be particularly important to insects. Proline has been shown to stimulate feeding and growth in many insects,[74,129] and Karolewski and Pucacki[107] found that both drought and frost stress as well as fumigations with SO_2 increased the levels of proline in young Norway spruce. The proline content correlated closely with the number of necrotic flecks on the fumigated trees. Tesche[183] found the same effects, and combinations of several stresses revealed that the proline accumulation works in an additive way. The exact function of proline in stressed plants is unknown but it seems to play a role in osmotic regulation and in chlorophyll synthesis.[41]

Glutathione (a tripeptide formed from glutamine, cysteine, and glycine) is able to stimulate both insect feeding and growth. Glutathione is known to play a role as scavenger of H_2O_2 in the chloroplast of plants, while in insects it has been shown to take part in the detoxification of some naturally occurring antibiotics and toxins.[30] The already-mentioned stimulating effect of SO_2 on the growth of Mexican bean beetles on soybean correlated closely with the content of glutathionine in the leaves, with an increase in glutathione at SO_2 concentrations from 50 to 300 ppb for 24 h. (ibid.) The probable role of glutathione in the response was demonstrated in a later experiment where Mexican bean beetles were fed excised soybean leaves that had been allowed to imbibe a glutathione solution. The beetles showed a significant preference for the glutathione-treated leaves, and larval growth, survival, and rate of development were improved by the treatment.[96] On the other hand, growth, rate of development, and survival of the cabbage looper *Trichoplusia ni* was unaffected by glutathione when reared on GSH-augmented soybean leaves.[95]

German field studies have shown increases in glutathione in needles of SO_2-damaged Norway spruce. The glutathione content was increased two- to fivefold in the damaged trees compared to nonexposed trees.[71,73] In addition, the damaged trees showed increases in cysteine, another S-containing amino acid. Smidt[176] found increases in SH compounds of *Picea abies* after short-term exposure to high concentrations of SO_2 and O_3 (0.25 to 2 ppm for both gases). In this connection it is worth noting the correlation found on spruce in Austria between the attack of *Sacchiphantes* sp. and total sulfur[165] and between the attack by the sawfly *Pristiphora abietina* and the SH content of the needles.[12] Whether glutathione and other SH-containing nitrogen compounds have a general stimulatory effect on insects, however, remains to be shown.

C. WATER CONTENT OF THE FOOD

There is some evidence that the water content of the food is of great importance to some insect species and that low water content can reduce the efficiency of nitrogen uptake.[130] The effects of air pollutants on the water content of leaves and needles have not been investigated in much detail, but the increases in soluble sugars and amino acids may change the osmotic potential of the cells and thus the water content of the tissues. Significant increases in water content have been observed in *Phleum pratense* fumigated with 90 ppb SO_2 and 100 ppb NO_2 for 19 d[127] and in Norway spruce subjected to NO_2[103] or a combination of SO_2, NO_2, and O_3.[110]

D. MINERALS AND SALTS

It seems inevitable that changes in the mineral content of plants could lead to changes in the behavior and success of insect herbivores, but the present stage of knowledge on the mineral requirements of insects makes it difficult to draw any general conclusions. The general notion is that insects require rather high levels of potassium, phosphate, and magnesium while their need for calcium, sodium, and chloride is limited.[39] High contents of calcium, sodium, and potassium have been shown to reduce the feeding activity of some insects and the same has been seen for the anions sulphate, iodine, and carbonate.[27]

Changes in the mineral content of plants under the influence of acid rain and air pollution have been observed in many cases. A common response in forest trees is reductions in Mg, Ca, and K and increased levels of S, F, Pb, Zn, Cu, Fe, Mn, and Al in the foliage.[153] The changes are caused both by a considerable leaking of nutrients from the foliage and by changing uptake from the soil. Flückiger et al.[57] found that increases in nitrogen and decreases in potassium in weakened beech stands in Schweitz correlated with increased attack by the aphid *Phyllaphis phagi* and the beetle *Rhyncaenus fagi.*

Salts deposited on the plant surface can significantly change the feeding preference of insects. Spraying Bermuda grass with 14 different salts reduced the feeding of larvae of fall army worm *Spodoptera frugiperda* on all sprayed plants compared to controls.[117] One of the applied salts was ammonium sulphate, an important constituent of acid precipitation. It thus seems most likely that dry or wet deposited salts, heavy metals, etc. could change the behavior of insects in the field, and perhaps the tendency for increases in sucking and mining species and decreases in many surface-feeding species is connected to some antifeeding property of deposited materials.

E. VITAMINS AND STEROLS

The vitamin requirements of insects are not quite clear, but in general seven B-complex vitamins seem to be essential, although some species may receive the needed compounds from symbiotes in the gut. Not all insects need vitamin C in the food, but there is evidence that insects feeding on plants high in ascorbic acid have lost the capacity of synthesis. Vitamin C has been shown to be essential to seven species of lepidopterans[39] and experiments with the lepidopteran *Laspeyresia pomonella* have shown a positive correlation between the level of ascorbic acid in the diet and the fecundity of the moths.[175]

Changes in the vitamin content of plants under air pollution stress have been shown, but the number of experiments is limited and only some vitamins have been measured. Mehlhorn et al.[134] found increases in vitamins C and E in needles of *Picea abies* and *Abies alba* after long-term exposure to low levels of SO_2 and O_3 (12 ppb SO_2 and 37 ppb O_3). Similar increases in vitamin C were found in five pine species exposed to 50 ppb O_3 for 3 weeks, while the level was unchanged in trees fumigated with 150 ppb

O_3.[7] In apparent contrast to these investigations, Flückiger and Oertli[55] found decreasing levels of vitamin C in birch and hawthorn with increasing proximity to a motorway. The reason may be that the levels of O_3 will often be reduced close to a road due to reactions with NO, and thus the need for the antioxidant ascorbic acid may be limited here. Ozone treatment also affects the vitamin content of crop plants. Corn and cabbage showed significant increases in vitamin C, while a reduction was seen in tomato. The content of thiamine increased in cabbage and lettuce but decreased in tomato. Increases in the levels of niacin have shown in carrot and strawberry.[173] There is, however, no information on whether these latter changes are relevant to insect nutrition.

The metabolism of insects is remarkable in the inability to synthesize sterols, and they are therefore wholly dependent on sterols obtained through the food or via symbiotes.[9,39] Not much is known about air pollution effects on sterols, but changes in the sterol metabolism have been shown for bean plants.[140] Changes in vitamins and sterols in plants is expected to have the greatest effect on insects that do not depend on symbiotes. It is worth noting that all aphids possess endosymbiotes in their gut,[9] which may be part of the reason why aphids only rarely react negatively to air pollution.

F. SECONDARY METABOLITES

Secondary plant metabolites influence all levels of the plant-insect interaction, but work on changes in secondary compounds due to air pollution is quite limited and strongly biased towards effects on terpenoids and phenolics. The following section therefore addresses these two groups in the most detail.

1. Terpenoids and Other Volatiles

Terpenoids constitute a major part of the resin of conifers, which appears to be an important defense against insects and pathogens[119] and may also play a role in host selection.[25,125] Attack by the bark beetle *Ips typographus* on conifers, for instance, is countered by a primary flow of resin that often kills the insects, followed by a secondary flow that envelops the injured area and prevents fungal infections.[31]

Insects specialized in feeding on conifers have often developed detoxification systems and may even be able to use at least the monoterpenes. *Dendrolimus pini* has thus been shown to metabolize more than 90% of the ingested terpenes.[142] Systems of detoxification must have some metabolic cost, however, and early instars are often more strongly inhibited than later instars. The composition of the terpene pool is also important. American studies have shown that the relation between camphene and betapinene in provenances of Douglas fir seems to determine the mortality of the spruce budworm *Choristoneura occidentalis.*[145] The general role of resins is probably an effective defense against generalist herbivores, while their effect on specialized coniferophagous insects may be limited. The effects are expected to be greatest on the early instar larvae.

Few air pollution studies have focused on resins and terpenes, but they all suggest that gases may cause changes in their metabolism. Studies of ozone-injured ponderosa pine in California showed that these trees had reduced content and lower flow-rate of resin than healthy trees,[178] while the terpene composition seemed unchanged.[33] Changes in the resin content of needles were found as well. Cobb et al.[34] found that oxidant-injured pines had significantly reduced total content of resin in the needles compared to healthy trees. The difference was largest in 11-month-old needles, smaller in 8-month-old needles, and negligible in the 1-month-old needles. Almost all the variation in resin content could be attributed to reduced levels of the terpene methyl clavicol. Finnish field studies around two polluting factories showed differences in response. Around a fertilizer factory emitting SO_2, NO_2, NH_3, and HF there was an increase in volatile oil in needles of Scotch pine with increasing injury of the trees, whereas a negative correlation between injury class and resin content was found around an SO_2-emitting pulp mill. The total amounts of terpenes increased at the fertilizer plant and the authors suggest that this may be due to a stimulating effect of nitrogen deposition. The terpene composition was changed in injured plants. Increases in *n*-hexanal were observed at both sites, while camphene increased at the fertilizer plant and 3-carene increased at the paper mill.[116] Dässler[42] found reductions in the total resin content of Norway spruce needles after high-level fumigations with SO_2 (up to 10 ppm!).

The emission of volatiles active in insect host-finding are also changed by air pollutants. Renwick and Potter[155] found that short-term fumigation of *Abies balsamea* with high levels of SO_2 (2 ppm) increased the emission of a number of terpenes. The increased emission appeared before any visible injuries were recorded. Bucher[25] reported similar effects from long-term fumigations of Scotch pine with 225 ppb SO_2, and the increased emission coincided with a decrease in the terpene content of the needles. In this case, increased emission was seen only after the appearance of necrotic flecks on the needles.

General conclusions on the effect of such changes are difficult to draw due to the dual effects of terpenes as repelling/attracting and stimulating/inhibiting growth, but a hypothesis may be that an increased emission of terpenes could enhance the host-location of specialists, while generalist herbivores may be repelled. On the other hand, the generalists will probably be enhanced by reductions of the terpene/resin content of needles. It is thus possible that changes in the terpene metabolism could be a driving force in a changed composition of the herbivore fauna.

Increased release of ethylene and ethane is a common effect of air pollutants. Bucher[24] demonstrated that emission of ethylene increased from larch, Scotch pine, and Norway spruce when the plants were fumigated with SO_2 at 50 to 200 ppb, and Peiser and Yang[144] saw increases in the release of ethylene and ethane from sulfur-dioxide-injured alfalfa plants. Short-term exposure of soybean and tomato to high concentrations of ozone (750 to 950 ppb) gave increases in ethylene emission as well.[158] These volatiles may be important in the host-choice of insects as well. Ethane is known to be a strong repellant to the wood-boring beetle *Monochamus alternatus.*[181]

2. Phenolics

Simple phenolics are found in almost all plants, and the recognition of specific phenolics plays a major role in the host-plant selection by many insects.[88,118] Reductions in growth rate and increases in development times of insects have been found to correlate with increasing content of phenolics.[104,154] Homopterans are particularly sensitive to phenolics, and negative correlations between phenolic content of leaves and susceptibility to aphids have been shown in several instances.[119,188,201] However, there are also examples of improved growth as a result of increases in phenolics (gypsy moth, tree locust, silkworm, etc.).

A special group of phenolics are the tannins, which some authors believe influence protein uptake and general metabolic functions in insects either by complexing with the proteins of the diet[49] or by inhibiting proteolytic and other enzymes in the saliva or in the digestive tract.[70,198] These effects have probably never been shown *in vivo,* and the responses of insects to tannins are very complex.[13,118,202]

A number of studies have shown that the phenolic metabolism of plants can be influenced by air pollutants or acid rain. Howell[92] found elevated levels of caffeic acid in the leaves of *Phaseolus vulgaris* exposed to ambient air containing $<$110 ppb ozone compared to plants from a coal-filtered control. Hurwitz et al.[102] found increases in 4′7-dihydroxyflavone in leaves of alfalfa exposed to short-term fumigations with 200 or 300 ppb ozone, while Jones and Pell[106] found no changes in isoflavonoids of alfalfa treated similarly.

Changes in total phenolics have been demonstrated in a number of experiments. Tingey et al.[187] found increases in total phenolics in the needles but decreases in the roots of ponderosa pine seedlings exposed to 100 ppb ozone for 20 weeks. Yee-Meiler[199] fumigated young clones of *Picea abies, Alnus glutinosa,* and *Betula pendula* to 50, 100, and 200 ppb SO_2 for several months. Increases in the content of total phenolics and *o*-diphenols were seen in the fumigated *P. abies* throughout the experiment, while no change could be measured in the phenoloxidase activity. The fumigated *A. glutinosa* showed no change in total phenolics, but increased activity of phenoloxidase was registered immediately prior to the appearance of necrotic flecks on the leaves. In *B. pendula* neither the phenolic content nor the activity of phenoloxidase was affected. Tingey[186] observed signs of changes in the phenolic metabolism of bean plants exposed to 500 ppb ozone. The activity of phenyl-alanine-ammonium lyase (a key enzyme in the early steps of phenol synthesis) increased slightly, while the activity of polyphenol oxidase (PPO) increased strongly. PPO oxidizes simple phenolics to highly reactive quinones which can condense with other quinones to form large, irregular structures or react with other cell compounds such as amino acids, amines, and SH-containing proteins.[93]

Changes in the phenolic metabolism have been observed in the field as well. Analysis of SO_2-damaged needles of *P. abies* revealed that the damaged needles had increased levels of total phenolics, while the content of the simple phenol *p*-hydroxyacetophenon and its glucoside picein was diminished, compared to healthy needles, indicating changes in the composition of the phenolic compounds.[72] Increases in total phenolics were also observed in ozone-injured tobacco plants from the field compared to healthy plants.[135] In contrast to these observations, *B. pendula* at the verge of a motorway had decreased levels of total phenolics compared to trees at a distance of 40 m from the motorway.[56]

Virtually no work has been done on the content of tannins in relation to air pollution, although changes in tannins may be particularly important to chewing and mining insects. All the mentioned measurements have relied on rather unspecific methods that only reflect the number of phenolic groups in the material. A change in the phenolic pool from tannins to simple phenolics would be registered by these methods as an increase in total phenolics, because tannins condense via the phenolic groups of the simple phenolics. Schopf et al.[163] have found that monomeric anthocyanidins give a stronger response in the common colorimetric tests than an equivalent amount of polymeric anthocyanidins. Very little is known of the dynamics of tannins, but Haslam[79] has hypothesized that tannins may serve as a store from which simple phenolics can be mobilized in response to bacterial or fungal infections. One Danish experiment has shown that tannins may be reduced by air pollutants. A nonsignificant reduction in both tannins and total phenolics was observed in *P. abies* exposed to 150 ppb O_3 for 2 months, while trees treated with acid rain of pH 3.5 or 2.4 showed decreases in tannin of 4 to 5% ($p = 0.07$). The total phenolic content increased significantly in the pH 2.4 treatment ($p = 0.009$) and was unchanged at pH 3.4.[110] The tannin was measured as complex binding with hemoglobin. In contrast, French studies of visibly injured Norway spruce showed increases in tannins and flavonoids compared to healthy-looking trees, while no changes in the content of simple phenolics were observed. Tannins were isolated by complexing with gelatin.[36] Changes of this sort may be very significant to insects and need to be investigated in more detail.

3. Other Secondary Metabolites

Very little is known about effects of air pollution on other important groups of metabolites with known effect on herbivores (alkaloids, cyanogenic compounds, saponins, etc.). The only example is a report of decreased levels of alkaloids in the leaves of oxidant-injured tobacco plants compared to healthy plants.[135] There is thus a great need to investigate the response of some of these compounds in order to clarify the mechanisms behind the observed changes of insect herbivory in the field in polluted areas.

VIII. OTHER ENVIRONMENTAL CHANGES

Changes in the light intensity in forest canopies caused by leaf loss or abnormal shoot production can also lead to changes in the insect communities. For example, the sawfly *Cephalcia abietis* favors increased light, while the lepidopteran *Lymantria dispar* usually prefers more closed stands.[149]

The physical and chemical properties of the leaf surfaces may be altered by both gases and acid rain. Such changes may very well affect the ovipositional and feeding behavior of insects, but the experimental evidence for this is lacking. Leakage of amino acids, ions, and carbohydrates; deterioration of the cuticular wax layer; and changes in the surface pH have often been seen on leaf surfaces under the influence of air pollution.[177]

A number of experiments have shown that air pollution can lead to profound changes in the phylloplane flora of plants. Abnormal algal growth has, for example, been observed in many parts of Europe,[139] and Khanna[109] found big changes in the phylloplane microflora of *Croton bonplandianum* and *Musa sapientum* growing in an area polluted by a number of factories compared to a nonpolluted site. In general, the number of actinomycetes and fungi was higher, while a reduction of the bacterial number was recorded. Phylloplane flora could be important in insect-plant interactions as shown, for example, by Ellis et al.[47] who showed that the cabbage root fly *Della radicum* laid three to four times more eggs on radish plants with a rich phylloplane flora than on plants grown from seed treated with an antimicrobial agent. Reinoculation of the treated plants resulted in normal ovipositional behavior. Flückiger and Flückiger-Keller[54] found increases in the number of the bark-sucking coccid *Cryptococcus fagi* on beech in the area around Basel and suggest that increased algal growth on the trunks due to leaching of nutrients from the canopy may form a protective layer for them.

A further mechanism of change may be acidification and altered decomposition in the soil,[4,10] which may influence the survival of overwintering pupae in the litter and topsoil, but very little information is available on this topic. Führer[62] has demonstrated, however, that the mortality of prepupae of *Epinotia tedella* is highly influenced by the composition of the soil water and especially its content of phenolic compounds.

IX. CONCLUDING REMARKS AND FUTURE PROSPECTS

As the preceeding data have demonstrated, there is strong evidence of changes in insect herbivory induced by air pollution, but the understanding of the ecological significance of such changes is still limited. Hardly anything is known about the effects on plant and ecosystem productivity. A number of field studies suggest a simplification of the trophic systems, possibly leading to faster turnover of matter and less stability of the systems.[38,108] Most of the experiments performed, however, have only looked at the response of the insects and usually only at some indices of performance, which is not enough to permit conclusions on the system level. Hughes et al.[100] found that Mexican bean beetles damaged 45% of the leaf area of control plants, while 75% of the leaf area was damaged on plants treated with 135 ppb SO_2. Warrington et al.[193] have shown that an increase in aphid populations (*Acyrthosiphon pisum*) on *Pisum sativum* fumigated with 50 or 80 ppb SO_2 resulted in a 10% reduction in the yield of peas compared to aphid-infested control plants. These are the only studies demonstrating an effect of decreased productivity of a plant due to herbivory induced by air pollution, and much more work is needed to draw conclusions on the ecological and economic significance of the interactions.

Even less is known about long-term effects on population dynamics, as few experiments have looked at more than one or a few generations of insects. Observations from the field suggest at least two different types of change. One is the occurrence of mass outbreaks of hitherto harmless insects, and the other is a change in the frequency and extent of outbreaks of species with cyclic patterns of eruption. Insects naturally regulated at low levels have been coined "stealthy species" and are partly regulated by low nutritional value of the host plants.[157] Stealthy species have a preference for highly predictable tissues, perennials, and trees and for older tissues with generally low nutrient levels and high contents of tannins and lignin.[156] If air pollution has a dual effect of reducing parasitism and predation and increasing the general suitability of the plants to the insects, this may explain the release of a number of species from innocuous levels.

A number of species with cyclic eruptions seem to be switching between stealthy and opportunistic behavior. Rhoades[157] suggests that induced responses and their subsequent relaxation may play an important role in the cyclic variations, as these cannot be explained satisfactorily by changes in weather and natural enemies alone. A number of authors have reported changes in leaf defensive chemistry or nutritional value in response to herbivory or mechanical damage of tissues,[81,83,185] and some of these changes can last for several years.[80,82,164] Although it is highly disputed whether the effects are large enough to change population dynamics and whether they constitute a truly induced defense against herbivores or just a passive effect of nutrient loss and changed metabolism,[58] the theory warrants attention in respect to long-term effects of air pollution. If plants under air-pollution stress respond differently to herbivory, (for example, by exhibiting a weaker or less persistent induced response), this may lead to changes in the frequency and extent of insect outbreaks. This calls for detailed studies of the long-term development in the frequency of attack by herbivores in the field, and for experimental work over longer periods of time in which changes in response to herbivory and air pollution can be monitored from year to year. Finally, there seems to be a need for more studies combining fumigation experiments and diet experiments in order to clarify the biochemical basis of changes in insect growth and reproduction under air-pollution stress.

REFERENCES

1. **Alfaro, R. I. et al.,** Pine oil, a feeding deterent for the white pine weevil, *Pissodes strobi* (Coleoptera: Curculionidae), *Can. Entomol.,* 116, 41, 1984.
2. **Alstad, D. N. et al.,** Effects of air pollutants on insect populations, *Annu. Rev. Entomol.,* 27, 369, 1982.
3. **Augustine, M. G. et al.,** Host-plant selection by the Mexican bean beetle, *Epilachna varivestris, Ann. Entomol. Soc. Am.,* 57, 127, 1964.
4. **Båat, E. et al.,** Effects of artificial acid rain on microbial activity and biomass, *Bull. Environ. Contam. Toxicol.,* 23, 737, 1979.
5. **Baltensweiler, W.,** Waldsterben: forest pests and air pollution, *Z. Angew. Entomol.,* 99(1), 77, 1985.
6. **Barannik, A. P.,** Ecological faunistic characteristics of den drophilous entomofauna of green plantings in industrial cities of Kemorovo province, *Sov. J. Ecol.,* 10(1), 1979 (English transl.).
7. **Barnes, R. L.,** Effects of chronic exposure to ozone on soluble sugar and ascorbic acid

contents of pine seedlings, *Can. J. Bot.*, 50, 215, 1972.

8. **Beard, R. L.**, Observations on house flies in high-ozone environments, *Ann. Entomol. Soc. Am.*, 58(3), 404, 1965.
9. **Beck, S. D. and Reese, J. C.**, Insect-plant interactions: nutrition and metabolism, in *Biochemical Interactions Between Plants and Insects. Recent Advances in Phytochemistry,* Vol. 10, Wallace, J. W. and Mansell, R. L., Eds., Plenum Press, New York, 1975.
10. **Berg, B.**, The influence of experimental acidification on nutrient release and decomposition rate of needle and root litter in the forest floor, *For. Ecol. Manage.*, 15(3), 181, 1986.
11. **Berge, H.**, Beziehungen zwischen Baumschädlingen und Immissionen, *Anz. Schäedlingskd. Pflanz. Umweltschutz,* 46(10), 155, 1973.
12. **Berger, R. and Führer, E.**, Die Kleine Fichtenblattwespe in Hausruck: Immissionsökologische Untersuchungen zum Massenwechsel, in *Tagung der Fachgruppe Forst und Holzwirtschaft,* Universität für Bodenkultur Wein, Semin. 1, October 22 and 23, 1987, 25.
13. **Bernays, E. A.**, Plant tannins and insect herbivores: an appraisal, *Ecol. Entomol.*, 6, 353, 1981.
14. **Bink, F. A.**, Acid stress in *Rumex hydrolapathum* (Polygonaceae) and its influence on the phytophage *Lycaena dispar* (Lepidoptera, Lycaenidae), *Oecologia,* 70, 447, 1986.
15. **Bishop, J. A. et al.**, Moths, lichens and air pollution along a transect from Manchester to North Wales, *J. Appl. Ecol.*, 12(1), 83, 1975.
16. **Blais, J. R.**, Trends in the frequency, extent and severity of spruce budworm outbreaks in Eastern Canada, *Can. J. For. Res.*, 13, 539, 1983.
17. **Bolsinger, M. and Flückiger, M.**, Effect of air pollution at a motorway on the infestation of *Virbunum opulus* by *Aphis fabae, Eur. J. For. Pathol.*, 14, 256, 1984.
18. **Bolsinger, M. and Flückiger, W.**, Enhanced aphid infestation at motorways: the role of ambient air pollution, *Entomol. Exp. Appl.*, 45, 237, 1987.
19. **Bösener, R.**, Zum vorkommen rindenbrutender Schadinsekten in rauchgeschadigten Kiefern und Fichtenbestanden, *Arch. Forstwes.*, 18(9/10), 1021, 1969.
20. **Braun, S. et al.**, Einfluss einer Autobahn auf die gehalte an IES, Chlorophyll, RNS und protein bei *Betula pendula* und *Cornus sanguinea, Eur. J. For. Pathol.*, 10, 378, 1980.
21. **Braun, S. and Flückiger, W.**, Increased population of the aphid *Aphis pomi* at a motorway. I. Field evaluation, *Environ. Pollut. Ser. A,* 33, 107, 1984.
22. **Braun, S. and Flückiger, W.**, Population of the aphid *Aphis pomi* at a motorway. II. The effect of draught and deicing salt, *Environ. Pollut. Ser. A,* 36, 261, 1984.
23. **Braun, S. and Flückiger, W.**, Population of the aphid *Aphis pomi* at a motorway. III. The effect of exhaust gases, *Environ. Pollut. Ser. A,* 39, 183, 1985.
24. **Bucher, J. B.**, SO_2-induced ethylene evolution of forest tree foliage, and its potential use as a stress-indicator, *Eur. J. For. Pathol.,* 11, 369, 1981.
25. **Bucher, J. B.**, Emissions of volatiles from plants under air pollution stress, in *Gaseous Air Pollutants and Plant Metabolism,* Koziol, M. J. and Whatley, F. R., Eds., London, 1984.
26. **Cates, R. G. et al.**, Natural product defensive chemistry of Douglas-fir, western spruce budworm success and forest management practices, *Z. Angew. Entomol.,* 96, 173, 1983.
27. **Chapman, R. F.**, The chemical inhibition of feeding by phytophagous insects. A review, *Bull. Entomol. Res.*, 64, 339, 1974.
28. **Chappelka, A. H. et al.**, Effects of ozone on soybean resistance to the Mexican bean beetle (*Epilachna varivestris* Mulsant), *Environ. Exp. Bot.*, in press.
29. **Charles, P. J. and Villemant, C.**, Modifications des niveaux de population d'insectes dans les jeunes plantations de pins sylvestres de la foret de Roumare (Seine-Maritime) soumises a la pollution atmosphérique. *C. R. Acad. Agric. Fr.*, 63, 502, 1977.
30. **Chiment, J. J. et al.**, Glutathione as an indicator of SO_2-induced stress in soybean, *Environ. Exp. Bot.*, 26(2), 147, 1986.
31. **Christiansen, E. and Horntvedt, R.**, Combined *Ips/ce ratocystis* attack on Norway spruce, and defensive mechanisms of the trees, *Z. Angew. Entomol.*, 96, 110, 1983.
32. **Clausen, I. H. S.**, Notes on the impact of air pollution (SO_2 and Pb) on spider (Araneae) populations in North Zealand, Denmark, *Entomol. Medd.*, 52, 33, 1984.
33. **Cobb, F. W., Jr., et al.**, IV. Theory of the relationship between oxidant injury and bark beetle infestation, *Hilgardia,* 39(6), 141, 1968.
34. **Cobb, F. W., Jr., et al.**, Effect of air pollution on the volatile oil from leaves of *Pinus ponderosa, Phytochemistry,* 11, 1815, 1972.
35. **Coleman, J. S. and Jones, G. C.**, Acute ozone stress on eastern cottonwood (*Populus deltoides* Bartr.) and the pest potential of the aphid, *Chaitophorus populicola* Thomas (Momoptera: Aphididae), *Environ. Entomol.*, 17(2), 207, 1988.
36. **Contour-Ansel, D. and Lougnet, P.**, Variation du taux de polyphenols dans les

aguilles d'epiceas *(Picea abies)*, presentants differents degres de deperissement, *Pollut. Atmos.*, October-December, 270, 1986.

37. **Cramer, H. H.,** De geographischen Grundlagen des Massenwechsels von *Epiblema tedella* Cl., *Forstwiss. Centralbl.*, 70, 42, 1951.
38. **Dabrowska-Prot, E.,** Ecological analysis of *Diptera* communities in the agricultural region of the Masurian Lakeland and the industrial region of Silesia, *Pol. Ecol. Stud.*, 6(4), 685, 1980.
39. **Dadd, R. H.,** Insect nutrition: current developments and metabolic implications, *Annu. Rev. Entomol.*, 18, 281, 1973.
40. **Dahlsten, D. L. and Rowney, D. L.,** Influence of Air Pollution on Population Dynamics of Forest Insects and on Tree Mortality, Gen. Tech. Rep. PSW-43, USDA Forest Service, Berkeley, CA, 1980, 125.
41. **Dashek, W. V. and Erickson, S. S.,** Isolation, assay, biosynthesis, metabolism, uptake and translocation, and function of proline in plant cells and tissues, *Bot. Rev.*, 47(3), 349, 1981.
42. **Dässler, H. G.,** Der Einfluss des Schwefeldioxids auf den Terpengehalt von Fichtennadeln, *Flora,* 154, 376, 1964.
43. **Dewey, J. F.,** Accumulation of fluorides by insects near an emission source in western Montana, *Environ. Entomol.*, 2(2), 163, 1973.
44. **Dohmen, G. P.,** Secondary effects of air pollution: enhanced aphid growth, *Environ. Pollut. Ser. A,* 39, 227, 1985.
45. **Dohmen, G. P. et al.,** Air pollution increases *Aphis fabae* pest potential, *Nature (London),* 307, 52, 1984.
46. **Donaubauer, E.,** Durch Industrieabgase bedingte Sekundarschäden am Wald, *Mitt. Forstl. Bundes Versuchsanst. Mariabrunn,* 73, 101, 1965.
47. **Ellis, P. R. et al.,** The role of microorganisms colonising radish seedlings in the ovipositional behaviour of the cabbage root fly, *Delia radicum, in Proc. 5th Int. Symp. Insect-Plant Relationships,* Pudoc, Wageningen, The Netherlands, 1982, 131.
48. **Endress, A. G. and Post, S. L.,** Altered feeding preference of Mexican bean beetle, *Epilachna varivestris,* for ozonated soybean foliage, *Environ. Pollut. Ser. A,* 39, 9, 1985.
49. **Feeny, P. P.,** Inhibitory effects of oak leaf tannins on the hydrolysis of proteins by trypsin, *Phytochemistry,* 8, 2219, 1969.
50. **Feir, D. and Hale, R.,** Responses of the large milkweed bug, *Oncopeltus fasciatus* (Hemiptera: Lygaedidae), to high levels of air pollutants, *Int. J. Environ. Stud.*, 20, 269, 1983.
51. **Feir, D. and Hale, R.,** Growth and reproduction of an insect model in controlled mixtures of air pollutants, *Int. J. Environ. Stud.*, 20, 223, 1983.
52. **Fischer, P. and Führer, E.,** Nematoden als Natürliche Gegenspieler der Fichtenblattwespe *Cephalcia* sp. im Böhmerwald, in *Tagung der Fachgruppe Forst- und Holzwirtschaft,* Universität Für Bodenkultur Wien, October 22 and 23, 1987, 27.
53. **Flückiger, W. and Braun, S.,** Effects of Air Pollutants on Insects and Hostplant/Insect Relationships, in proc. of a workshop jointly organized within the framework of the Concerted Action: Effects of Air Pollution on Terrestrial and Aquatic Ecosystems. Working Party III: How are the Effects of Air Pollutants on Agricultural Crops Influenced by the Interaction with Other Limiting Factors?, Commission of the European Communities and the National Agency of Environmental Protection, Risö National Laboratory, Denmark. March 23 to 25, 1986.
54. **Flückiger, W. and Flückiger-Keller, H.,** Untersuchungen uber Waldschäden in der Nordwestschweiz, *Schweiz. Z. Forstwes.*, 135(5), 389, 1984.
55. **Flückiger, W. and Oertli, J. J.,** Observations of an aphid infestation on Hawthorn in the vicinity of a motorway, *Naturwissenschaften,* 65, 654, 1978.
56. **Flückiger, W. et al.,** Biochemische Veränderungen in jungen Birken im Nahbereich einer Autobahn, *Eur. J. For. Pathol.*, 8, 154, 1978.
57. **Flückiger, W. et al.,** Untersuchungen über Waldschäden in festen Beobachtungsflächen der Kantone Basel-Landschaft, Basel-Stadt, Aargau, Solothurn, Bern, Zürich und Zug, *Schweiz. Z. Forstwes.*, 137(11), 917, 1986.
58. **Fowler, S. V. and Lawton, J. H.,** Rapidly induced defenses and talking trees: the devil's advocate position, *Am. Nat.*, 126(2), 181, 1985.
59. **Friend, W. G.,** Nutritional requirements of phytophagous insects, *Annu. Rev. Entomol.*, 3, 57, 1958.
60. **Führer, E.,** Studien über den Fichtennestwickler *Epiblema (Epinotia) tedella* Cl. in Norddeutschland, *Anz. Schaedlingsk.*, 36(8), 118, 1963.
61. **Führer, E.,** Untersuchungen über die Ursachen der Befallsdisposition der europäischen Fichte gegenüber *Epibla tedella* Cl. (Lep.: Tortricidae), *Z. Angew. Entomol.*, 59, 292, 1967.
62. **Führer, E.,** Uber die parasitar bedingte, differenzierte Mortalität bei uberwinterden Larven von *Epiblema tedella* Cl. (Lep.: Tortricidae), *Z. Angew. Entomol.*, 69, 368, 1971.
63. **Führer, E.,** Das Immissionsproblem und der Forstschutz, *Allg. Forstzg.*, 94(7), 163, 1983.

64. **Führer, E.**, Air pollution and the incidence of forest insect problems, *Z. Angew. Entomol.*, 99(4), 371, 1985.
65. **Galecka, B.**, Structure and functioning of community of Coccinellidae (Coleoptera) in industrial and agricultural-forest regions, *Pol. Ecol. Stud.*, 6(4), 717, 1980.
66. **Galecka, B.**, Phenological development of *Frangula alnus* Mill. in an industrial region and the number of *Aphis frangulae,* Kalt., *Pol Ecol. Stud.*, 10(1-2), 141, 1984.
67. **Gilbert, O.**, Some indirect effects of air pollution on barkliving invertebrates, *J. Appl. Ecol.*, 8, 77, 1970.
68. **Ginevan, M. E. and Lane, D. D.**, Effects of sulphur dioxide in the air on the fruit fly, *Drosophila melanogaster, Environ. Sci. Technol.*, 12, 828, 1978.
69. **Ginevan, M. E. et al.**, Ambient air concentrations of sulphur dioxide affects flight activity in bees, *Proc. Natl. Acad. Sci. U.S.A.*, 77(10), 5631, 1980.
70. **Goldstein, J. L. and Swain, T.**, The inhibition of enzymes by tannins, *Phytochemistry,* 4, 185, 1965.
71. **Grill, D. and Esterbauer, H.**, Cystein und Glutathion in gesunder und SO_2-geschädigten Fichtennadeln, *Eur. J. For. Pathol.*, 3, 65, 1973.
72. **Grill, D. et al.**, Untersuchungen an phenolischen Substanzen und Glucose in SO_2-geschädigten Fichtennadeln, *Phytopathol. Z.*, 82, 182, 1975.
73. **Grill, D. et al.**, Effect of sulphur-dioxide on protein-SH in needles of *Picea abies, Eur. J. For. Pathol.*, 10, 263, 1980.
74. **Haglund, B. M.**, Proline and valine — cues which stimulate grasshopper herbivory during drought stress?, *Nature (London),* 288(18), 697, 1980.
75. **Hain, F. P.**, Interactions of insects, trees and air pollutants, *Tree Physiol.*, 3, 93, 1987.
76. **Hain, F. P. and Arthur, F. H.**, The role of atmospheric deposition in the latitudinal variation of Fraser fir mortality caused by the balsam woolly adelgid, *Adelges picea* (Rats.) (Hemipt., adelgidae): a hypothesis, *Z. Angew. Entomol.*, 99(2), 145, 1985.
77. **Hall, R. W. et al.**, Effects of atmospheric deposition treatments on suitability of elms for elm leaf bettle *(Xanthogaleruca luteola)* (Coleoptera: Chrysomelidae), *Environ. Entomol.*, in press.
78. **Harborne, J. B.**, *Introduction to Ecological Biochemistry,* 2nd ed., Academic Press, New York, 1980.
79. **Haslam, E.**, Vegetable tannins, in *Biochemistry of Plant Phenolics* (Recent Advances in Plant Physiol., 12), Swain, T. et al., Eds., Plenum Press, New York, 1979, 474.
80. **Haukioja, E.**, On the role of plant defences in the fluctuation of herbivore populations, *Oikos,* 35, 202, 1980.

80a. **Haukioja, E.**, personal communication.
81. **Haukioja, E. and Hanhimäki, S.**, Rapid wound-induced resistance in white birch *(Vetula pubescens)* foliage to the geometrid *Eppirita autumnata:* a comparison of trees and moths within and outside the outbreak range of the moth, *Oecologia,* 65, 223, 1985.
82. **Haukioja, E. and Neuvonen, S.**, Induced long-term resistance of birch foliage against defoliators: defensive or incidental?, *Ecology,* 66(4), 1303, 1985.
83. **Haukioja, E. and Niemelä, P.**, Retarded growth of a goemetrid larva after mechanical damage to leaves of its host tree, *Ann. Zool. Fenn.*, 14, 48, 1977.
84. **Heliövaara, K.**, Occurrence of *Petrova resinella* (Lepidoptera, Tortricidae) in a gradient of industrial air pollutants, *Silva Fenn.*, 20(2), 83, 1986.
85. **Heliövaara, K. and Väisänen, R.**, Industrial pollution and the pine bark bug, *Aradus cinnamomeus* (Het. Aradidae), *J. Appl. Entomol.*, 101, 469, 1986.
86. **Heliövaara, K. and Väisänen, R.**, Parasitization in *Petrova resinella* (Lepidoptera, Tortricidae) galls in relation to industrial air pollutants, *Silva Fenn.*, 20(3), 233, 1986.
87. **Heliövaara, K. et al.**, Parasitism in the eggs of the pine bark-bug *Aradus cinnamomeus* (Heteroptera, aradidae), *Ann. Entomol. Fenn.*, 48(1), 31, 1982.
88. **Heron, R. J.**, The role of chemotactic stimuli in the feeding behavior of spruce budworm larvae on white spruce, *Can. J. Zool.*, 43, 247, 1965.
89. **Hillman, R. C.**, Biological Effects of Air Pollution on Insects, Emphasizing the Reaction of the Honeybee to Sulphur Dioxide, Ph.D. thesis, Pennsylvania State University, University Park, 1972.
90. **Hillmann, R. C. and Benton, A. W.**, Biological effects of air pollution on insects, emphasizing the reactions of the honeybee (*Apis Mellifera* L.) to sulphur dioxide, *J. Elisha Mitchell Sci. Soc.*, 88, 195, 1972.
91. **House, H. L.**, Effects of different proportions of nutrients on insects, *Entomol. Exp. Appl.*, 12, 651, 1969.
92. **Howell, R. K.**, Influence of air pollution on quantities of caffeic acid isolated from leaves of *Phaseolus vulgaris, Phytopathology,* 60, 1626, 1970.
93. **Howell, R. K.**, Phenols, ozone and their involvement in pigmentation and physiology of plant injury, in *Air Pollution Effects on Plant Growth,* (ACS Symposium Series, No. 3), Dugger, M., Ed., American Chemical Society, Washington, D.C., 1974.
94. **Hughes, P. R.**, Effects of air pollution on

plant-feeding insects, Doc. No. 84/85, in Proc. 12th Annu. ENR Conf., Urbana-Champaign, IL, September 13 to 14, 1983.

95. **Hughes, P. R.,** Insect populations on host plants subjected to air pollution, in *Plant Stress — Insect Interactions,* Heinrichs, E. A., Ed., John Wiley & Sons, in press.

96. **Hughes, P. R. and Chiment, J. J.,** Enhanced success of the Mexican bean beetle (Coleoptera: Coccinellidae) on glutathione-enriched soybean leaves, *Environ. Entomol.,* preprint.

97. **Hughes, P. R. et al.,** Effects of air pollutants on plant-insect interactions: reactions of the Mexican bean beetle to SO_2-fumigated pinto beans, *Entomol. Soc. Am.,* 10(5), 741, 1981.

98. **Hughes, P. R. et al.,** Effects of air pollution on plant-insect interactions: increased susceptibility of greenhouse-grown soybeans to the Mexican bean bettle after plant exposure to SO_2, *Environ. Entomol.,* 11(1), 173, 1982.

99. **Hughes, P. R. et al.,** Modification of insect-plant relations by plant stress, in *Proc. 5th Int. Symp. Insect-Plant Relationships,* Pudoc, Wageningen, The Netherlands, 1982.

100. **Hughes, P. R. et al.,** Increased success of the Mexican bean beetle on field-grown soybeans exposed to sulphur dioxide, *J. Environ. Qual.,* 12(4), 565, 1983.

101. **Hughes, P. R. et al.,** Effect of pollutant dose on the response of Mexican bean beetle (Coleoptera: Coccinellidae) to SO_2-induced changes in soybean, *Environ. Entomol.,* 14(6), 175, 1985.

102. **Hurwitz, B. et al.,** Status of coumestrol and 4′7-dihydroxyflavone in alfalfa foliage exposed to ozone, *Phytopathology,* 69(8), 810, 1979.

103. **Huttunen, S. et al.,** Changes in osmotic potential and some related physiological variables in needles of polluted Norway spruce *(Picea abies), Ann. Bot. Fenn.,* 18, 63, 1978.

104. **Isman, M. B. and Duffey, S. S.,** Phenolic compounds in foliage of commercial tomato cultivars as growth inhibitors to the fruitworm *Heliothis zea, J. Am. Soc. Hortic. Sci.,* 107(1), 167, 1982.

105. **Jeffords, M. R. and Endress, A. G.,** Possible role of ozone in the defoliation by the gypsy moth (Lepidoptera: Lymantriidae), *Environ. Entomol.,* 13(5), 1249, 1984.

106. **Jones, J. V. and Pell, E. J.,** The influence of ozone on the presence of isoflavones in alfalfa foliage, *J. Air Pollut. Control Assoc.,* 31(8), 885, 1981.

107. **Karolewski, P. and Paukacki, P.,** Sensitivity of polish races of Norway spruce (*Picea abies* (L.) Karst) to the action of sulphur dioxide and low temperatures, *Abor. Kórnickie,* 28, 129, 1983.

108. **Katayev, O. A. et al.,** Changes in arthropod communities of forest biocoenoses with atmospheric pollution, *Entomol. Rev.,* 62, 20, 1983.

109. **Khanna, K. K.,** Phyllosphere microflora in certain plants in relation to air pollution, *Environ. Pollut. Ser. A,* 42, 191, 1986.

110. **Kielberg, L. et al.,** Luftforurening og Insektangreb Pà Rödgran, M.Sc. thesis, Institute of Plant Ecology, University of Copenhagen, Denmark, 1987.

111. **Koziol, M. J.,** Interactions of gaseous pollutants with carbohydrate metabolism, in *Gaseous Air Pollutants and Plant Metabolism,* Koziol, M. J. and Whatley, F. R., Eds., London, 1984.

112. **Larsson, S. et al.,** Responses of *Neodiprion sertifer* (Hym. Diprionidae) larvae to variation in needle resin acid concentration in Scots pine, *Oecologia (Berlin),* 70, 77, 1986.

113. **Lebrun, P. et al.,** Tests écologiques de toxitolerance au SO_2 sur l'oribate corticole *Humerobates rostrolamellatus* (Granjean 1936), (Acari: Oribatei), *Ann. Soc. R. Zool. Belg.,* 106, 193, 1976.

114. **Lechowicz, M. J.,** Resource allocation by plants under air pollution stress: implications for plant-pest interactions, *Bot. Rev.,* 53(3), 281, 1987.

115. **Leetham, J. L. et al.,** Responses of heterotrophs, in *The Effects of SO_2 on Grassland. Ecological Studies No. 45,* Lauenroth, W. K. and Preston, E. M., Eds., Springer-Verlag, Berlin, 1984.

116. **Lehtiö, H.,** Effect of air pollution on the volatile oil in needles of Scots pine (*Pinus sylvestris* L.), *Silva Fenn.,* 15(2), 122, 1981.

117. **Leuck, D. B. et al.,** Nutritional plant sprays: effect on fall armyworm feeding preferences, *J. Econ. Entomol.,* 67(1), 58, 1974.

118. **Levin, D. A.,** Plant phenolics: an ecological perspective, *Am. Nat.,* 105(942), 157, 1971.

119. **Levin, D. A.,** The chemical defenses of plants to pathogens and herbivores, *Ann. Rev. Ecol. Syst.,* 7, 121, 1976.

120. **Levy, R. et al.,** Effects of ozone on three species of diptera, *Environ. Entomol.,* 1(5), 608, 1972.

121. **Levy, R. et al.,** Tolerance of three species of insects to prolonged exposures to ozone, *Environ. Entomol.,* 3(1), 184, 1972.

122. **Loomis, W. D. and Bataille, J.,** Plant phenolic compounds and the isolation of plant enzymes, *Phytochemistry,* 5, 423, 1966.

123. **Ludwig, W. and Liebig, G.,** Die Populationsdynamik der Kleinen Fichtenquirlschildlaus Physokermes hemicryphus (Dalman) in verschieden Stark erkrankten Fichtenbestän den, in Projekt Europäisches Forschungszentrum für Massnahmen zur

Luftreinhaltung, 2. Statuskolloquim des PEF vom 4. bis 7. März 1986 in Kernforschungszen trum Karlsruhe, KfK-PEF 4, 1986, 359.

124. **Mader, H. J.,** Die Isolationswirkung von Verkehrsstrassen auf Tierpopulationen untersucht am Beispiel von Arthropoden und Kleinsäugern der Waldbiozoenose. Schriftenr. Landschaftspflege und Naturschuta, Heft 19, Bonn-Bad Godersberg, 1979.

125. **Maziara-Borusiewicz, K.,** *Zeiraphera diniana Gn.* (Lepidoptera, Tortricidae) and the chemical composition of the needles of larch (*Larix decidua* Mill.) from the Alps and from the Polish mountains, *Z. Angew. Entomol.,* 95, 414, 1983.

126. **Malhotra, S. S. and Sarkar, S. K.,** Effects of sulfur dioxide on sugar and free amino acid content of pine seedlings, *Physiol. Plant.,* 47, 223, 1979.

127. **Mansfield, T. A. et al.,** Interactions between the responses of plants to pollution and other environmental factors such as drought, light and temperature, in Proceedings of a workshop jointly organized within the framework of the Concerted Action: Effects of Air Pollution on Terrestrial and Aquatic Ecosystems, Working Party III: How are the Effects of Air Pollutants on Agricultural Crops Influenced by the Interaction with Other Limiting Factors?, Commission of the European Communities and the National Agency of Environmental Protection, Risö National Laboratory, Denmark, March 23 to 25, 1986.

128. **Mattson, W. J.,** Herbivory in relation to plant nitrogen content, *Ann. Rev. Ecol. Syst.,* 11, 119, 1980.

129. **Mattson, W. J. and Haack, R. A.,** The role of drought in outbreaks of plant-eating insects, *Bioscience,* 37(2), 110, 1987.

130. **Mattson, W. J. and Scriber, J. M.,** Nutritional ecology of insect folivores of woody plants: nitrogen, fiber and mineral consideration, in *Nutritional Ecology of Insects, Mites, Spiders and Related Invertebrates,* Slansky, F. and Rodrigues, J. G., Eds., 1987.

131. **McNary, T. J. et al.,** Effect of controlled low levels of SO_2 on grasshopper densities on a northern mixed grass prairie, *Entomol. Soc. Am.,* 74(1), 91, 1981.

132. **McNeill, S. and Southwood, T. R. E.,** The role of nitrogen in the development of insect/plant relationships, in *Biochemical Aspects of Plant and Animal Coevolution,* Harborne, J. B., Ed., Academic Press, New York, 1978.

133. **McNeill, S. and Southwood, T. R. E.,** The interaction between air pollution and sucking insects, in *Acid Rain: Scientific and Technical Advances,* Terry, R. et al., Eds., Selper Ltd., London, 1987, 602.

134. **Mehlhorn, H. et al.,** Effects of SO_2 and O_3 on production of anti-oxidants in conifers, *Plant Physiol.,* 82, 336, 1986.

135. **Menser, H. A. and Chaplin, J. F.,** Air pollution: effects on the phenol and alkaloid content of cured tobacco leaves, *Tob. Sci.,* 169, 73, 1969.

136. **Miller, P. L.,** Oxidant-induced community change in a mixed conifer forest, in *Air Pollution Damage to Vegetation,* Advances in Chemistry Series No. 122, Naegele, J. A., Ed., American Chemical Society, Washington, D.C., 1973.

137. **Mitterböck, F. and Führer, E.,** Wirkungen fluorbelasteter Fichtennadeln auf Nonnenraupen *Lymantria monacha* L. (Lep.: Lymantriidae), *Z. Angew. Entomol.,* in press.

138. **Neuvonen, S. and Lindgren, M.,** The effects of simulated acid rain on performance of the aphid *Euceraphis betulae* (Koch) on silver birch, *Oecologia,* 74, 77, 1987.

139. **Nihlgard, B.,** The ammonium hypothesis — an additional explanation to the forest dieback in Europe, *Ambio,* 14(1), 3, 1985.

140. **Ormrod, E. P.,** *Pollution in Horticulture. Fundamental Aspects of Pollution Control and Environmental Science 4,* Elsevier, Amsterdam, 1978.

141. **Otto, D.,** Zur bedeutung des Zuckergehaltes der Nahrung für die Entwicklung nadelfressender Kieferninsek ten, *Arch. Forstwes.,* 19, 135, 1970.

142. **Otto, D.,** Zur bedeutung des Diefernnadelharzes und des Kiefernnadelöles für die Entwicklung nadel fressender Insekten, *Arch. Forstwes.,* 19, 151, 1970.

143. **Parry, W. H.,** The effects of nutrition and density on the production of alate *Elatobium abietinum* on Sitka spruce, *Oecologia,* 30, 377, 1977.

144. **Peiser, G. G. and Yang, S. F.,** Ethylene and ethane production from sulfur dioxide-injured plants, *Plant Physiol.,* 63, 142, 1979.

145. **Perry, D. A. and Pitman, G. B.,** Genetic and environmental influences in host resistance to herbivores: Douglas-fir and the western spruce budworm, *Z. Angew. Entomol.,* 96, 217, 1983.

146. **Petal, J.,** The effects of industrial pollution of Silesia on populations of ants, *Pol. Ecol. Stud.,* 6(4), 665, 1980.

147. **Petters, R. M. and Mettus, R. V.,** Reproductive performance of *Bracon hevetor* females following acute exposure to sulphur dioxide in air, *Environ. Pollut. Ser. A,* 27, 155, 1982.

148. **Pfeffer, A.,** Insektenschädlinge an Tannen in Bereich der Gasexhalation, *Z. Angew. Entomol.,* 51, 203, 1963.

149. **Pfeffer, A.,** Wirkungen von Luftverunreinigungen auf die freilebende Tierwelt, *Schweiz. Z. Forstwes.,* 129(1), 362, 1978.

150. **Port, G. R. and Thompson, J. R.,** Outbreaks of insect herbivores on plants along motorways in the United Kingdom, *J. Appl.*

Ecol., 17, 649, 1980.

151. **Prestridge, R. A. and McNeill, S.**, The role of nitrogen in the ecology of grassland *Auchenorryncha*, in *Nitrogen as an Ecological Factor*, Lee, J. A. et al., Eds., Blackwell, Palo Alto, CA, 1983, chap. 12.
152. **Przybylski, Z.**, The effects of automobile exhuast gasses on the arthropods of cultivated plants, meadows, and orchards, *Environ. Pollut.*, 19, 157, 1979.
153. **Rademacher, P.**, Morphologische und physiologische Eigenschaften von Fichten (*Picea abies* L. Karst.), Tannen (*Abies alba* Mill.), Kiefern (*Pinus sylvestris* L.) und Buche (*Fagus sylvatica* L.) gesunder und er kranker Waldstandorte, Diss. vom Fachbereich Biologie der Universität Hamburg, GSSK-Forschungszen trum Geesthacht, GmbH Geesthacht, 1986.
154. **Reese, J. C. and Beck, S. D.**, Effects of allelochemics on the black cutworm, *Agrotis ipsilon*, effects of cathecol, L-dopa, dopamine and chlorogenic acid on larval growth, development and utilization of food, *Ann. Entomol. Soc. Am.*, 69(1), 59, 1976.
155. **Renwick, J. A. A. and Potter, J.**, Effects of sulfur dioxide on volatile terpene emission from balsam fir, *J. Air Pollut. Control Assoc.*, 31, 65, 1981.
156. **Rhoades, D. F.**, Toward a general theory of plant antiherbivore chemistry, in Biochemical Interactions Between Plants and Insects, (Recent Advances in Phytochemistry, Vol. 10), Wallace, J. W. and Mansell, R. L., Eds., Plenum Press, New York, 1975, 168.
157. **Rhoades, D. F.**, Offensive-defensive interactions between herbivores and plants: their relevance in herbivore population dynamics and ecological theory, *Am. Nat.*, 125(2), 205, 1985.
158. **Rodecap, K. D. and Tingey, D. T.**, Ozone induced ethylene release from leaf surfaces, *Plant Sci.*, 44, 73, 1986.
159. **Rosen, P. M. and Runeckles, V. C.**, Interactions of ozone and greenhouse whitefly in plant injury, *Environ. Conserv.*, 3(1) 70, 1976.
160. **Ruszcyk, A.**, Mortality of *Papilio scamander scamander* (Lep., Papilionidae) pupae in four districts of Porto Alegre (S. Brazil) and the causes of the superabundance of some butterflies in urban areas, *Rev. Bras. Biol.*, 46(3), 567, 1986.
161. **Schopf, R.**, Zur Bedeutung des inhaltstoffmusters von fichtennadeln für die Entwicklung phytophager Insekten, *Z. Angew. Entomol.*, 96, 166, 1983.
162. **Schopf, R.**, Zur Nahrungsqualität von Fichtennadeln für Forstliche Schadinsekten. XX. Korrelationen der Konzentrationen von Fichtennadelninhaltstoffen mit Entwicklungsparametern der Blattwespe *Gilpinia hercyniae* Htg. (Hym. Diprionidae), *Z. Angew. Entomol.*, 95, 189, 1983.
163. **Schopf, R. et al.**, As to the food quality of spruce needles for forest damaging insects. XVIII. Resorption of secondary plant metabolites by the sawfly *Gilpinia hercynia* Htg. (Hym., Diprionidae), *Z. Angew. Entomol.*, 93, 244, 1982.
164. **Schultz, J. A. and Baldwin, I. T.**, Oak leaf quality declines in response to defoliation by gypsy moth larvae, *Science*, 217, 149, 1982.
165. **Schwaninger, C. and Führer, E.**, Immissionsökologische Untersuchungen an Fichtenläusen (Adelgidae) im Bereich des mittleren Inntales, in Forschungsinitiative gegen des Waldsterben. Bericht 1987, Ergebnisse aus der Immissionsforschung; Bundesministerium für Wissenschaft und Forschung, Wien, 1987, 195.
166. **Scriber, J. M.**, Host-plant suitability, in *Chemical Ecology of Insects*, Bell, W. J. and Cardé, R. T., Eds., London, 1984.
167. **Seitschek, O.**, Die gegenwärtige Waldschutz-Situation in Bayern, *Allg. Forstz.*, 37(15), 423, 1982.
168. **Sierpinski, Z.**, Schädliche Insekten an Jungen Kiefernbestanden in Rauchschadengebieten in Oberschlesien, *Arch. Forstwes.*, 15(10), 1105, 1966.
169. **Sierpinski, Z.**, Einfluss von industriellen Luftverunreinigungen auf die Populationsdynamik einiger primärer Kiefernschädlinge. XIV, IUFRO-Kongress, München 1967, Referate, Bind 5, Section 24.
170. **Sierpinski, Z.**, Pests of spruce forest ecosystems, in Stability of Spruce Forest Ecosystems, Proc. Symp. Brno, Czechoslovakia, October 29 to November 2, 1979, 335.
171. **Sierpinski, Z.**, Über den Enfluss von Luftberunreinigungen aus Schadinsekten in polnischen Nadelbaumbeständen, *Forstwiss. Centralbl.*, 103, 83, 1985.
172. **Sierpinski, Z.**, Luftverunreinigungen und Forstschädlinge, *Z. Angew. Entomol.*, 99, 1, 1985.
173. **Skärby, L.**, Changes in the nutritional quality of crops, in *Gaseous Air Pollutants and Plant Metabolism*, Koziol, M. J. and Whatley, F. R., Eds., London, 1984.
174. **Skuhravá, M. and Skuhravy, V.**, Outbreak of two gall midges, *Harrisomyian. gen. vitrina* (Kieffer) and *Drisina glutinosa* Giard. (Cecidomyiidae, Diptera) on maple, *Acer pseudoplatanus* L., in Czechoslovakia, with descriptions of the two genera and species, *J. Appl. Entomol.*, 101, 256, 1986.
175. **Smelyanets, V. P.**, Mechanisms of plant resitance in Scotch pine *(Pinus sylvestris)*. IV. Influence of food quality on physiological state of pine pests *(Trophic preferendum)*,

Z. Angew. Entomol., 84, 232, 1977.

176. **Smidt, S.,** Begasungsversuche mit SO_2 und Ozon an jungen Fichten, *Eur. J. For. Pathol.,* 14, 241, 1984.
177. **Smith, W. H.,** Air pollution and forests, in *Interactions between Air Contaminants and Forest Ecosystems,* Springer-Verlag, Berlin, 1981.
178. **Stark, R. W. and Cobb, F. W.,** Smog injury, root diseases and bark beetle damage in Ponderosa pine, *Calif. Agric.,* 23, 13, 1969.
179. **Stary, P. and Kubiznáková, J.,** Content and transfer of heavy metal air pollutants in populations of *Formica* spp. wood ants (Hym., Formicidae), *J. Appl. Entomol.,* 104, 1, 1987.
180. **Strojan, C. L.,** The impact of zinc smelter emissions on forest litter arthropods, *Oikos,* 31, 41, 1978.
181. **Sumimoto, M. et al.,** Ethane in pine needles preventing feeding of the beetle, *Monochamus alternatus, J. Insect Physiol.,* 21, 713, 1975.
182. **Templin, E.,** Zur Populationsdynamik einiger Kiefernschadinsedten in rauchgeschädigten Beständen, *Wiss. Z. Tech. Univ. Dresden,* 11(3), 631, 1962.
183. **Tesche, M.,** The influence of environmental stress upon spruce, in UNESCO/IUFRO: Stability of Spruce Forest Ecosystems, Proc. Symp., Brno, Czechoslovakia, October 29 to November 2, 1979.
184. **Thalenhorst, W.,** Untersuchungen über den Einfluss fluorhaltiger Abgase auf die Disposition der Fichte für den Befall durch die *Gallenlaus sacciphantes abietis, Z. Pflanzenkr.,* 82, 717, 1974.
185. **Thielges, B. A.,** Altered polyphenol metabolism in the foliage of *Pinus sylvestris* associated with European pine sawfly attack, *Can. J. Bot.,* 46, 724, 1968.
186. **Tingey, D. T.,** Ozone induced alterations in the metabolite pools and enzyme activities of plants, in *Air Pollution Effects on Plant Growth,* (ACS Symposoium Series No. 3, American Chemical Society, Washington, D.C., 1974.
187. **Tingey et al.,** The effect of chronic ozone exposures to the metabolite content of ponderosa pine seedlings, *For. Sci.,* 22(3) 234, 1976.
188. **Tjia, B. and Houston, D.,** Phenolic constituents of Norway spruce resistant or susceptible to the eastern spruce gall aphid, *For. Sci.,* 21(2), 180, 1975.
189. **Trumble, J. T. et al.,** Ozone-induced changes in host-plant suitability: interactions of *Keiferia lycopersicella* and *Lycopersicon esculentum, J. Chem. Ecol.,* 13(1), 203, 1987.
190. **Valentine, H. T. et al.,** Nutritional changes in host foliage during and after defoliation and their relation to the weight of gypsy moth pupae, *Oecologia,* 57, 298, 1983.
191. **Villemant, C.,** Influence de la pollution atmospherique sur les populations d'aphides du pin sylvestre en Foret de Roumaire (Seine Maritime), *Environ. Pollut. Ser. A,* 24, 245, 1981.
192. **Warrington, S.,** Relationship between SO_2 dose and growth of the pea aphid, *Acyrtosiphon pisum,* on peas, *Environ. Pollut. Ser. A,* 43, 155, 1987.
193. **Warrington, S. et al.,** Effect of SO_2 on the reproduction of pea aphids, *Acrytosiphon pisum* and the impact of SO_2 and aphids on the growth and yield of peas, *Environ. Pollut.,* 48, 285, 1987.

193a. **Warrington, S.,** personal communication.

194. **Wentzel, K. F.,** Insekten als Immissionsfolgeschädlinge, *Naturwissenschaften,* 52(5), 113, 1965.
195. **Wentzel, K. F.,** Ursachen des Waldsterbens in Mitteleuropa, *Allg. Forstz.,* 37(45), 1365, 1982.
196. **Wentzel, K. F. and Ohnesorge, B.,** Zum auftreten von Schadinsekten bei Luftverunreinigung, *Forstarchiv,* 32(9), 177, 1961.
197. **White, T. C. R.,** A hypothesis to explain outbreaks of Looper caterpillars, with special reference to populations of *Selidosema suavis* in a plantation of *Pinus radiata* in New Zealand, *Oecologia,* 16, 279, 1984.
198. **Wint, G. R. W.,** The effect of foliar nutrients upon the growth and feeding of a lepidopteran larva, in, Lee, J. A. et al., Eds., *Nitrogen as an Ecological Factor,* Blackwell, Palo Alto, CA, 1983.
199. **Yee-Meiler, D.,** Der Einfluss von kontinuierliche, nie drigen SO_2-Begasungen auf den Phenolgehalt und die Phenoloxidase-Aktivität in Blättern einiger Waldbaumarten, *Eur. J. For. Pathol.,* 8, 14, 1978.
200. **Zedler, B. et al.,** Impact of atmospheric pollution on the protein and amino acid metabolism of spruce *Picea abies* trees, *Environ. Pollut. Ser. A,* 40, 193, 1986.
201. **Zucker, W. V.,** How aphids choose leaves: the roles of phenolics in host selection by a galling aphid, *Ecology,* 63(4), 972, 1982.
202. **Zucker, W. V.,** Tannins: does structure determine function? An ecological perspective, *Am. Nat.,* 121(3), 335, 1983.

4 Extrinsic Factors Influencing Production of Secondary Metabolites in Plants

Peter G. Waterman
Phytochemistry Research Laboratories
Department of Pharmacy (Pharmaceutical Chemistry)
University of Strathclyde
Glasgow, Scotland

Simon Mole
Department of Biochemistry
Purdue University
West Lafayette, Indiana

TABLE OF CONTENTS

I. INTRODUCTION

Until recently, the ecological viewpoint on the occurrence and distribution of allelochemicals was shaped largely by the ideas of Feeny[33] and Rhoades and Cates.[110] Their interlinked hypotheses assumed the optimal allelochemical profile of a plant or plant part to be primarily determined by its dependability as a food source to herbivores. They predicted that in order to limit herbivore damage, common, large, and/or long-lived plants needed to make "costly" investment in high amounts of dose-dependent allelochemicals such as polyphenols (tannins) or resins. By contrast, rare, small, and/or short-lived plants (or plant parts) could escape from herbivores in space and time and optimum defense would be achieved by a "cheaper" investment in small amounts of qualitative (toxin) allelochemics.

Unfortunately, this simplistic view of what determines the secondary metabolism of a species cannot be sustained. An analysis of the distribution of secondary metabolites in the Angiospermae[72] indicates that with evolutionary advancement there has been a fundamental shift in emphasis from products of the shikimic acid (SA) pathway to those derived from acetate or mevalonate. This change closely follows the transition from the woody to the herbaceous habit[3] (a change largely coincident with the apparent-unapparent division recognized by Feeny[33]).

The taxonomic basis of secondary metabolite distribution still allows for allelochemic activity to be a major cause for production and proliferation through conservation and development of metabolic pathways that confer advantage. For the ecologist there is the particular problem of trying to understand the costs and benefits that are involved in producing these allelochemicals.[32] A hypothesis now gaining wide acceptance suggests that a major determinant of the optimum allelochemic output of a plant is the necessity to provide resources for production of the various precursors of secondary metabolism.[4,5,11,12,15,21] The proposition is that environmental stresses that influence the carbon/nutrient balance of a plant, when summed over a period, set the constraints that determine the cost-benefit relationships under which a plant operates and thus the suitability of different allelochemic strategies. The hypothesis has two distinct facets: (1) that based on the genotype, which governs where a taxon can survive and the range of allelochemic options open to it (i.e., the subset of metabolites that a plant could produce) and (2) that concerned with the phenotypic responses of an individual to changes in the environmental stresses it encounters. Evidence on the impact that extrinsic factors have on secondary metabolism at both levels is examined in this chapter, but with greater emphasis on phenotypic variation. Observations will then be rationalized on the basis of present understanding of the pathways and processes of secondary metabolism that occur in plants.

II. EXTRINSIC FACTORS INFLUENCING PRODUCTION OF SECONDARY METABOLITES

Evidence exists that a number of extrinsic factors do influence secondary metabolism.[89] The two most widely implicated variables are major soil nutrients and incident light intensity, and it is the interplay between these that is generally considered in discussions of resource allocation. In this section, studies monitoring the effects of soil nutrients and light intensity are dealt with separately and then the few which have dealt with both are considered. A number of other factors, including mechanical injury of the plant, influence metabolic processes and are also discussed.

This is not intended to be an exhaustive review of the literature, and only examples

Table 1
CHANGES IN THE CHEMISTRY OF THE LEAVES OF *BRASSICA NIGRA* GROWN UNDER LOW SULFUR (S) AND LOW NITROGEN (N) CONDITIONS[135]

			Weight of 7-d larvae	
Treatment	ITS	PROT	Pr	Se
Low S	35	80	183	594
Low N	140	67	159	67
Low N and S	78	70	94	29

Note: Concentrations of allylisothiocyanate (ITS) and protein (PROT) and weight of 7-d larvae of *Pieris rapae* (Pr) and *Spodoptera eridania* (Se) are expressed as a percentage of values for control plants and larvae.

considered to be most relevant are highlighted. Much work has been published on the influence of various extrinsic factors on metabolism of cell cultures,[29] but this is not considered. Such studies should be intepreted with caution in relation to whole plant metabolism, as compartmentation of biochemical processes can differ markedly between cell and differentiated organism, with consequences for the quantity and type of end-product produced.

A. MAJOR SOIL NUTRIENTS

In agriculture, addition of nutrients, particularly nitrogen, is often employed to increase biomass production. However, all plants do not react equally to an increased availability of nutrients; species genetically adapted to low-nutrient soils often fail to respond.[18] The impact of changes in nutrient availability on production of secondary metabolites has recently been reviewed.[40] The effects are to some extent unpredictable; trends can be recognized but it is not possible to make hard and fast rules.

Net production of nitrogenous metabolites (alkaloids, cyanogenic glycosides, and glucosinolates) is generally increased under nitrogen fertilizers. However, this increase may not keep pace with additional biomass production and as a consequence the actual concentrations found in tissues may decrease. It has been suggested[40] that the relative concentration of carbon to nitrogen in an alkaloid may be important in deciding if synthesis occurs in response to additional nitrogen. Low C to N ratios (4:1 to 16:1) often lead to increased alkaloid production; where ratios are higher results are less predictable, perhaps because the greater demands for additional carbon cannot be met. Phosphorus and potassium levels have variable effects on production of nitrogenous secondary metabolites. Lowering the level of phosphorus is reported to increase synthesis of cyanogenic glycosides, but this is not necessarily true for alkaloids.[40] Additional potassium can lower alkaloid production.[40] The potential impact of nutrient deficiency on the palatability of leaves of *Brassica nigra,* the major allelochemic of which is the glucosinolate aglycone allylisothiocyanate (I), has been demonstrated by Wolfson.[135] Reduction in available sulfur diminishes allelochemic production and there is an increase in the rate of development of the larvae of *Pieris rapae* and *Spodoptera eridania* fed on foliage from S-limited plants (Table 1). Surprisingly, lowering nitrogen availability increases isothiocyanate levels as well as reducing leaf nitrogen, but larvae of *P. rapae* still appear to perform better than on control plants. Use of nitrogen fertilizer diminishes glucosinolate production in *Isomeris arborea* to about half the concentration produced in plants treated with nitrogen-free fertilizer.[6]

$$CH_2{=}CHCH_2N{=}C{=}S$$

I

Nonnitrogenous metabolites derived from the shikimic acid pathway (simple cinnamic acids [II-V], hydrolysable tannins [VI], lignin, condensed tannins [VII] — in part) are generally produced in higher levels when plants are grown in nutrient-deficient conditions. Deficiencies in nitrogen, phosphorus, sulfur, and potassium have all been reported to cause increased production.[40] In acid soils there is a reduction in the rate of conversion of ammonium to nitrate which may reduce uptake of nitrogen in some species. Increasing soil acidity has been found to lead to higher levels of condensed tannin production in some tree species.[26] A comparison of the chemistry of foliage from common tree species growing in the relatively nutrient-rich soils of the Kibale Forest, Uganda and those from the highly acidic, nutrient-poor soils of the Douala-Edea Forest Reserve, Cameroon revealed relatively higher levels of phenolic compounds and greater lignification in the latter and relatively more protein nitrogen and alkaloids in the former (Table 2).[38,130] In the Korup National Park in Cameroon, species growing on soils particularly deficient in phosphorus produce foliage containing very high levels of polyphenols.[131] However, the proposition[59] that acidic, nutrient-poor soils are always associated with high levels of allelochemic production has not been confirmed by studies in lowland rain forests in Borneo.[1] At Kibale, Douala-Edea, and Korup the differences in secondary metabolite production appear to be largely at the genotypic level; between Douala-Edea and Kibale there is almost no species overlap, while between Douala-Edea and the somewhat less acidic conditions of Korup there is no clear distinction on levels of polyphenol production among over 30 species common to both sites.[129a] Similarly, it is reported[98] that polyphenol levels do not change significantly when *Copaifera multijuga* is grown on different soils in the Amazonian forest.

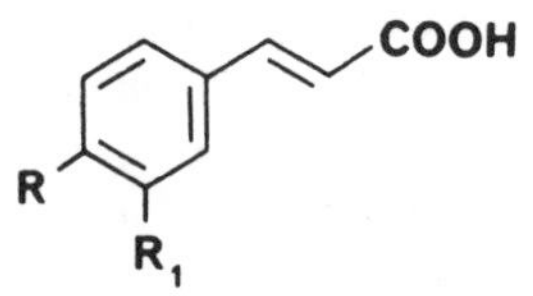

COOH

R

R_1

II CINNAMIC ACID $R = R_1 = H$

↓

III p-COUMARIC ACID $R = OH; R_1 = H$

↓

IV CAFFEIC ACID $R = R_1 = OH$

↓

V FERULIC ACID $R = OH; R_1 = OCH_3$

Table 2
A COMPARISON OF SOME CHEMICAL MEASURES FOR MATURE LEAVES FROM TREE SPECIES FROM THE KIBALE FOREST, UGANDA AND THE DOUALA-EDEA FOREST RESERVE, CAMEROON[131]

Site	PROT	ADF	CT	ALK
Kibale	16.5	34.6	1.3	48
Douala-Edea	11.0	55.2	4.2	16

Note: Protein (PROT), acid detergent fiber (ADF), and condensed tannins (CT) are expressed as weighted percentage values reflecting the relative biomass of species studied, and alkaloids (ALK) as the percentage of species studied that give strong positive to tests for this class of allelochemic.

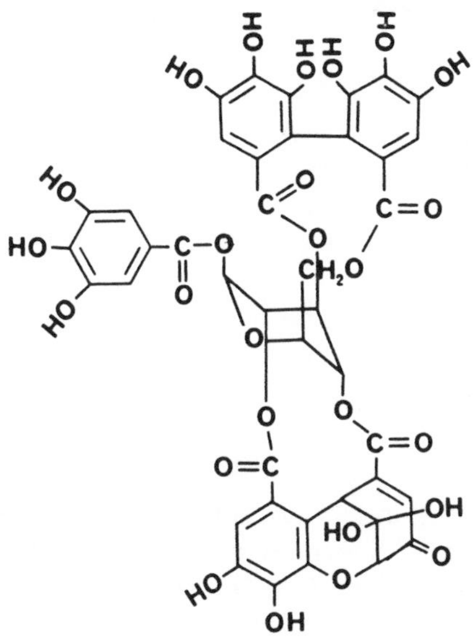

VI GERANIIN

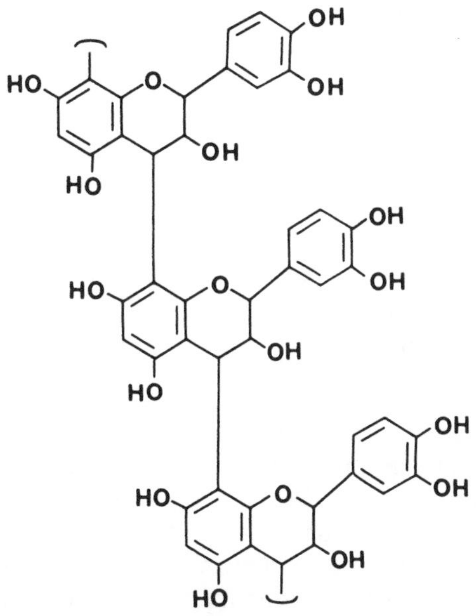

VII

Products of the mevalonic acid pathway (terpenes, e.g., VIII-X) do not show a consistent relationship to changes in nitrogen, phosphorus, or potassium nutrition.[40] Investigation[93] of the production of monoterpene and sesquiterpene metabolites by *Heterotheca subaxillaris* in conditions where available nitrate was variable revealed that as levels of NO_3^- were reduced from 15 to 0.5 m*M* terpene concentration rose from 3.1 to 5.1 mg/g. Feeding trials suggested that the increasing terpene levels in leaves, coupled with a lower nitrogen content, led to a decline in consumption, growth, and survival rate in larvae of *Pseudoplusia includens.*[92]

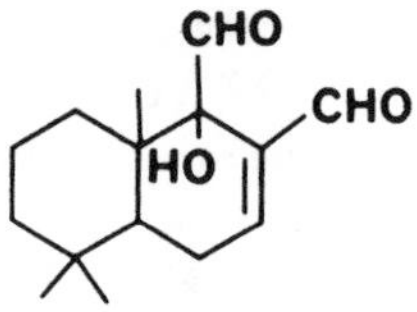

VIII SESQUITERPENE (WARBURGANAL)

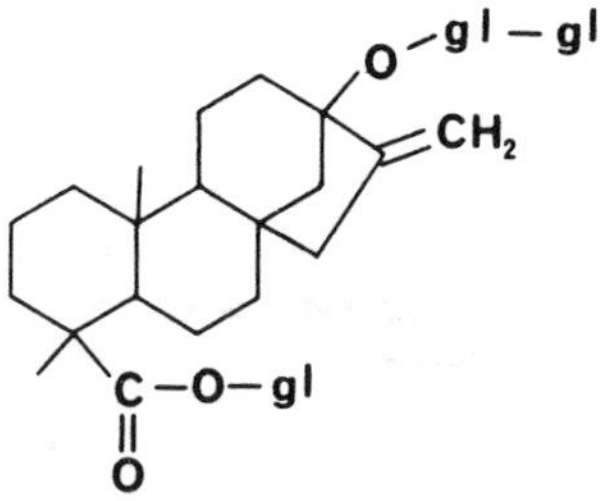

IX DITERPENE (STEVIOSIDE)

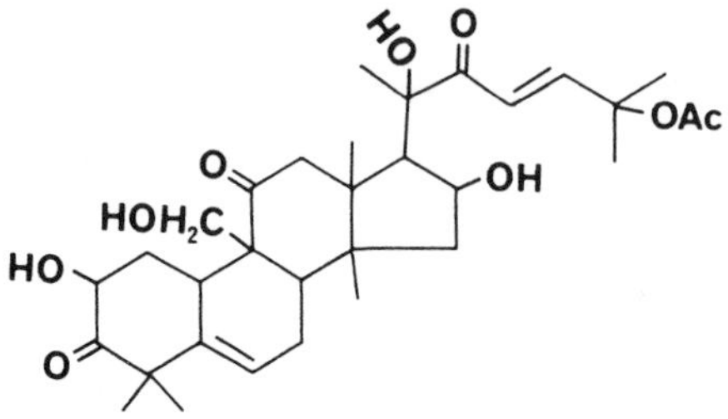

X TRITERPENE (CUCURBITACIN A)

B. LIGHT INTENSITY

Different taxa are adapted to enormous variation in incident light intensity. For example, between the exposed canopy and the deeply shaded understory of tropical rain forests there can be a 100-fold difference in incident light flux. Not surprisingly, there are distinctive differences in the photosynthetic apparatus of species adapted to sun and shade environments.[14,104,118] Growth patterns, such as increasing branching and leaf production[134] and the regeneration of stump suckers in birch (see Section II.G),[63] correlate with increasing light intensity. Nevertheless, while species may be adapted to low-light environments, it makes considerable intuitive sense that a plant that is shaded but otherwise not significantly stressed has reduced photosynthetic activity and fewer resources to contribute to secondary metabolism. However, it must also be remembered that high light intensity can often be associated with water stress and under such conditions metabolic processes may be limited.[90]

A number of studies[56,83,95,96,100,132,136] have related changes in incident light to production of phenolic allelochemics. Different light intensities in two growing seasons were presumed to have caused appreciable variation in phenolic content of *Sorghum bicolor* growing under identical soil and temperature conditions.[136] An analysis of foliage from individuals of the West African rain forest tree *Barteria fistulosa* growing in shaded and insolated locations showed differences in several measures related to secondary metabolic output (Table 3).[132] More recently, a study of four West African rain forest species has shown the correlation between incident light intensity and synthesis of all phenolics (including condensed tannins) to occur within as well as between individuals (Figure

Table 3
COMPARISON OF COMPOSITION OF MATURE LEAVES FROM INDIVIDUAL TREES OF *BARTERIA FISTULOSA* GROWING IN RELATIVELY HIGH-LIGHT (N = 15) AND LOW-LIGHT (N = 11) ENVIRONMENTS IN DOUALA-EDEA FOREST RESERVE, CAMEROON[132]

	TP	CT	ADF	CDIG	PROT
High light	8.46	19.10	45.90	28.24	10.50
Low light	5.39	11.96	45.59	35.04	12.13

Note: Values are averaged and given as percentage dry weight. Total phenolics = TP, condensed tannins = CT, acid detergent fiber = ADF, protein = PROT, and cellulase digestibility = CDIG.

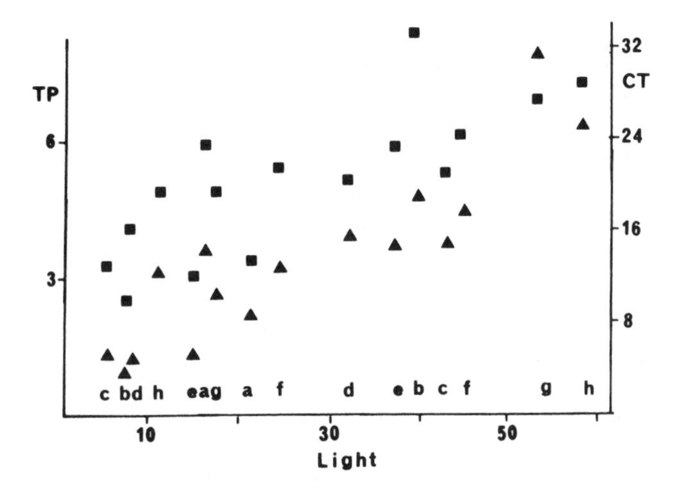

FIGURE 1. Total phenol (TP) and condensed tannin (CT) levels in leaves from *Acacia pennata* in different light environments. The data represent eight trees (a to h) from which two leaf samples of each were taken from relatively insolated and shaded branches. Concentrations of CT are denoted by triangles and of TP by squares for each sample. Light intensity is given as a percentage of total light flux measured in an open area. Values for TP and CT are as percentage dry weight relative to tannic acid and quebracho tannin, respectively.

1).[95] It is noteworthy that in neither of these studies[95,132] did it appear that lignification was occurring, suggesting that additional shikimate metabolism was directed specifically toward production of phenolics and tannins.

The influence of light is not restricted to phenolic compounds. Levels of both individual diterpene components and total diterpene resin in the needles of *Pinus sylvestris*[43] were higher in insolated than in shaded foliage, increasing by up to 100%. Shaded individuals of *P. taeda* produce less resin in response to attack by the southern pine beetle than do dominant or codominant individuals.[102]

The impact of light-induced changes in the levels of secondary metabolites produced may represent an important source of variation to herbivores.[43,56,83,95,96,101,132,136] The pattern emerging is that individuals within a species, or parts of an individual exposed to higher light intensities, are likely to respond by producing more secondary metabolites than congeners in more shaded environments. This is a phenotypic response. It is of interest that when interspecies (genotypic) comparisons are made in the rain forest biome the opposite trend is found. Those species that specialize in colonizing forest gaps tend to have lower levels of phenolics than do species of the climax forest.[19,130] Presumably, such colonizing species have a metabolism geared to a higher and more consistent supply of nutrient and photosynthate to rapid vegetative and reproductive development.

C. CARBON/NUTRIENT BALANCE

Substrate for secondary metabolism originates primarily from carbon fixation or, where nitrogen, sulfur, phosphorus, or halides are involved, from soil and/or waterborne nutrients. Of the nutrient-derived elements of secondary metabolism, nitrogen is by far the most widely incorporated. Utilization of sulfur occurs sporadically, but phosphorus is rarely employed which may suggest that it is generally in short supply.[21] Other elements become significant in environments where they occur in appreciable amounts, e.g., selenium halides in marine organisms.

The ability of both light flux and the level of soil nutrients to modify secondary metabolic output has been demonstrated (Sections II.A and II.B). There is substantial evidence that species adapted to environments where soils are of poor nutritional quality do tend to place a greater emphasis on production of wholly carbon-based secondary metabolites.[11,12,15,21,30] The resource allocation hypothesis rationalizes this by predicting that, even taking into acccount a maximum deployment of resources to nutrient capture by the root system,[15] there will remain an excess of photosynthate for conversion into nitrogen-free secondary metabolites. If this interpretation is correct, then carbon fixation and nutrient capture are not independent, and for a proper assessment of their impact on allelochemic productivity they must be considered in concert.

Larsson et al.[76] conducted a series of experiments in which leaf chemistry and palatability of genetically homogenous *Salix dasyclados* was studied in relation to manipulation of light and nutrient environments. Three growing conditions were employed: (1) an assumed nonstressed condition in which adequate nutrients were supplied at high light flux, (2) a nutrient-stressed state in which light intensity was unaltered but nutrient supply was restricted, and (3) a light-stressed state in which light intensity was less than 25% of (1) and (2) while nutrient levels were as in (1). Highest total phenol levels were observed in the optimized plants (1). Levels were slightly lower, but not significantly so, in the nutrient-stressed state (2) where it appeared that excess carbon was primarily allocated to root growth, presumably to assist with nutrient capture. Where plants were light stressed (3), phenolic output was severely reduced (Table 4). HPLC analysis of the foliage did not reveal qualitative differences in the compounds produced under the different treatments. The significance of these changes on leaf palatability was tested using *Galerucella lineola,* which showed a preference for leaves from light-stressed plants (Table 4).

Table 4
SOME CHANGES IN CHEMICAL AND PHYSIOLOGICAL CHARACTERISTICS AND IN PALATABILITY TO *GALERUCELLA LINEOLA* OF *SALIX DASYCLADOS* GROWN UNDER NUTRIENT- AND LIGHT-LIMITING CONDITIONS[76]

Measure	Light-limited	Nutrient-limited
Relative growth rate (% d^{-1})	50	44
Root dry wt/total dry wt	68	216
Nitrogen in leaves (% dry wt)	96	64
Total phenolics in leaves (% dry wt)	30	76
Palatability (mg dry wt eaten/plant)[a]	526	121

[a] Estimated from histograms.[76]

Note: Results are given as a percentage relative to levels found in control plants.

In a study of the chemistry of *Betula resinifera,* Bryant et al.[13] found that fertilizing the juvenile growth form led to increased herbivory of twigs by snowshoe hares. The greater palatability was correlated with a reduction in condensed tannins, total phenolics, and the antifeedant triterpene papyriferic acid (XI). While chemical defense was reduced by both N and NP treatments, P alone had the opposite effect and gave highest concentrations of phenolics (but not papyriferic acid) and minimum herbivory. Shading of control plants reduced levels of all chemical defenses and led to herbivory levels as high as those found in NP-fertilized individuals. Unfortunately, no combined shading and fertilizer studies were undertaken. In a comparable series of experiments on *Alnus crispa,* all fertilizer treatments produced general reduction in levels of phenolics and condensed tannins, the effect of N fertilizer on condensed tannin being particularly dramatic (approximately 80% reduction). Two other allelochemicals, pinosylvin (XII) and pinosylvin methyl ether, showed no reduction with N or P treatments but were reduced by combined N and P fertilizer. Palatability to snowshoe hares did not appear to be greatly modified by any treatment other than combined N and P fertilizer.

Bryant et al.[13] argued that the defense chemistry and palatability of the birch showed greater sensitivity to external factors than did that of the alder. They equated this with

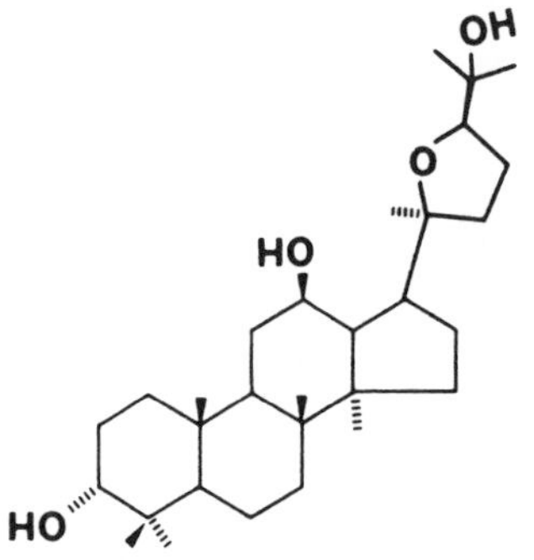

XI PAPYRIFERIC ACID

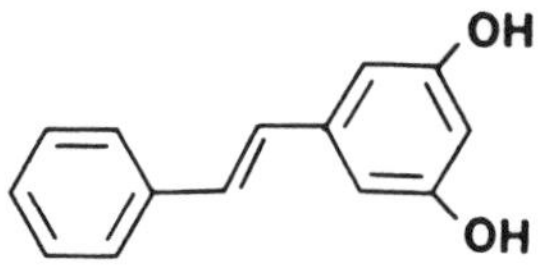

XII PINOSYLVIN

the fact that the former was adapted to conditions of relatively high resource availability, whereas the latter had a comparatively limited nutrient absorption capacity.[17] It seems to us that this contention is only partly sustained by the facts. From the plots presented, it appears that N fertilizer led to at least as great a relative reduction in the synthesis of phenolics, including condensed tannins, in alder as it did in birch. However, in the specific alder allelochemics pinosylvin and its methyl ether, there does seem to be less change than for the specific allelochemic of birch, papyriferic acid. Thus, if there are distinctions in the secondary metabolic changes that occur in the two species, they seem to relate to specific pathways and not to the ubiquitous simple phenolics and condensed tannins.

A comparison of the effects of shading and NP fertilizer on the chemistry of *S. alaxensis*[10] offers the best available comparison of their separate effects on a range of metabolites. Some of the results were predictable (Table 5); both treatments tended to cause a reduction of soluble leaf carbohydrate and in various estimates of phenolic content, and enhanced the palatability of twigs to snowshoe hares. While shading increased levels of nutrients, except soluble carbohydrates, NP fertilizers surprisingly caused a reduction in levels of all nutrients measured. Where fertilizer treatment was combined with shading, all nutrients except carbohydrate increased.

Carbohydrate storage, in the form of starch, can be modified by changes in nutrient availability and light-flux density. In *B. pendula,* concentrations of starch in the leaf were greatest at low nutrient levels and high light density and starch storage in roots occurred only in conditions of nutrient stress.[88] It was suggested that under nutrient-deficient conditions, available nitrogen was employed in maintaining photosynthesis rather than growth, which diminished.

D. SPECIFIC TRACE ELEMENTS (MICRONUTRIENTS)

There is relatively little information available on the impact of micronutrients on the production of secondary metabolites in whole plants. Lack of boron reduces the production of phenolic allelochemics by oil palm seedlings to the extent that they are more palatable to spider mites.[108] Tin and bismuth, but not iron, increase the quantity of tannin produced by *Acacia catuchu.*[69] It is not known whether these elements act directly on metabolism or cause a secondary response through toxicity.

E. OSMOTIC STRESSES (DROUGHT AND SALINITY)

The most common metabolic response to drought appears to be in the accumulation

Table 5
CHANGES IN BIOMASS PRODUCTION AND CHEMISTRY OF TWIGS OF *SALIX ALAXENSIS* GROWN EITHER SHADED OR TREATED WITH NP FERTILIZER[10]

Measure	Shaded	Fertilized
Biomass	−***	+***
Nitrogen	+***	−***
Phosphorus	+***	−***
Ether-soluble extractives	+*	−***
Soluble carbohydrate	−***	−*
Folin-Denis estimate of phenolics	−***	−***
Vanillin estimate of phenolics	−***	−***
Proanthocyanidin for condensed tannins	−***	−***
Astringency assay	−***	−***

Note: − indicates reduction and + indicates increases relative to controls; * indicates $p = 0.05$, ** indicates $p = 0.01$, and *** indicates $p = 0.001$.

of amino acids, notably proline, polyamines, sugars, cyclitols, betaine, choline, and inorganic ions.[87] Production of these compounds may be physiologically necessary to maintain osmotic balances; most can be viewed as potential nutrient sources and seem likely to increase palatability.

Drought conditions have been reported to lead to an increase (phenotypic) in several types of secondary metabolite: cyanogenic glycosides, glucosinolates, terpenoids, alkaloids, and condensed tannins.[9,40,87,105] Populations of *Isomeris arborea* growing in arid environments produce greater amounts of glucosinolates than populations in non-arid environments.[6] The Mediterranean species *Artemisia alba* produces an oil rich in sesquiterpene hydrocarbons when growing in mesophytic communities.[81] However, there is no consistent effect of drought stress; for example, there is no increase in alkaloid production by *Catharanthus roseus*.[37] The effect of drought on concentrations of metabolites is sometimes found to be dependent on the degree of stress and the time over which it occurs. Short-term effects may lead to increased production, whereas over the long term the opposite effect is observed.[87,90]

Salinity can be considered as a stress factor comparable to drought and again often leads to synthesis of osmoregulatory nitrogenous compounds, notably proline.[42] Increasing salinity does influence production of some secondary metabolites, including increases in simple phenolic acids[27] and alkaloids[8] and decreases in terpenes.[31]

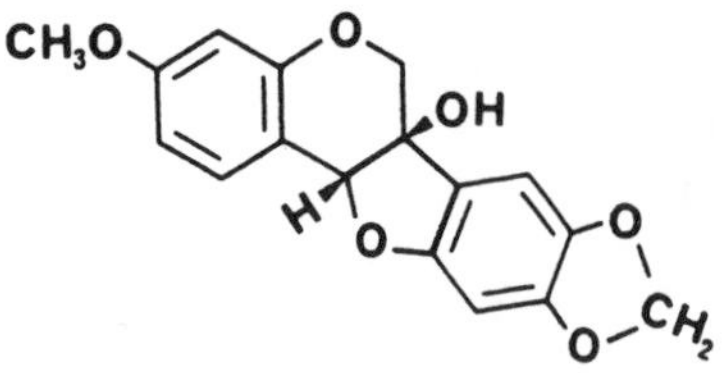

XIII PISATIN

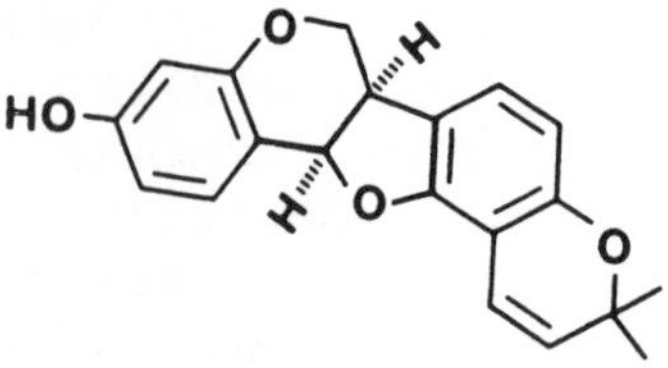

XIV PHASEOLLIN

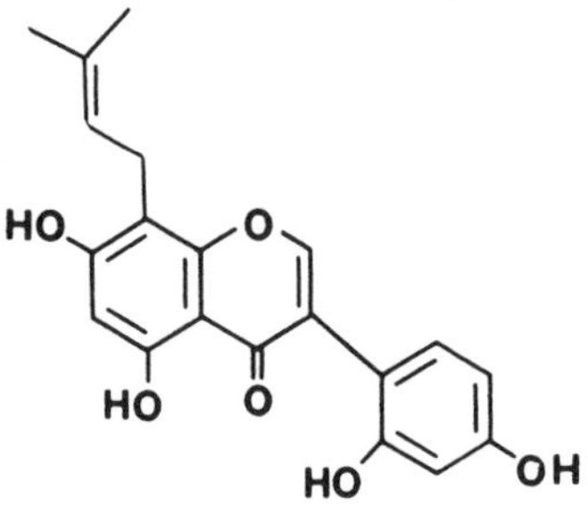

XV KIEVITONE

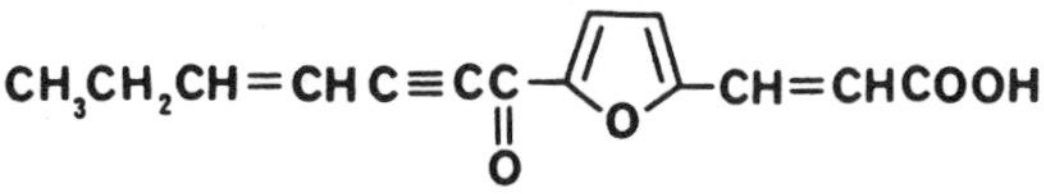

XVI WYERONE ACID

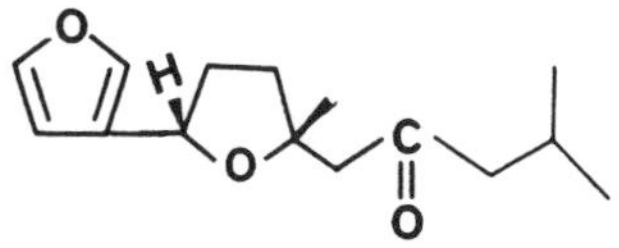

XVII IPOMEAMARONE

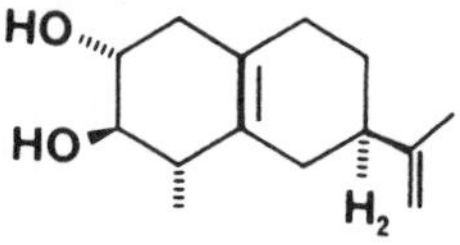

XVIII RISHITIN

FIGURE 2. Structures of some representative phytoalexins: XIII, XIV, and XV are flavonoid type (derived from SA and acetate); XVI is wholly acetate derived; and XVII and XVIII are formed from mevalonic acid.

F. INDUCED RESPONSES IN PLANT SECONDARY METABOLISM

Damage caused by wounding or pathogen invasion often leads to a biochemical response by the plant that reduces acceptability of the organ or the whole organism to future attackers. The clearest example of this is in the *de novo* production of phytoalexin substances in response to pathogen invasion.[30,73,125] Phytoalexin substances are commonly flavonoid derivatives, but both polyketide and terpenoid compounds are also formed (Figure 2). Proteinase-inhibitor proteins are rapidly produced by some plants in response to the activity of chewing insects, and this can significantly decrease the palatability of the plant or plant part.[70,114] The mechanism by which the synthesis of these substances is "switched on" appears to involve oligosaccharide or glycoprotein fragments (elicitors) derived from damaged plant cell walls or from the invading pathogens.[30,73,114]

Less-well-defined phenomena are short-term or long-term responses to wounding by increased production of phenolic compounds, in some cases identified as condensed tannins. Relatively small but reproducible enhancement in phenolics, sometimes only within the wounded organ and sometimes more widespread, has been found in a number of plants, for example in the *Betula* sp. in the U.K.[48,137] and Finland,[53,54] in cotton,[68] in feltleaf willow (condensed tannins specifically),[10] in Douglas fir (condensed tannins),[129] in sorghum[136] and in oak (increase in tanning ability[116] and decrease in nutrients[126]). Wounding can also induce increased synthesis of alkaloids, particularly in leaves that are actively growing,[37] and of monoterpenes.[77]

While wound-induced changes do certainly occur, the significance of the phenomenon remains difficult to assess.[35] It is by no means universal,[16,77,97] and changes that occur do not necessarily deter herbivores.[48] The level of induction seems to vary according to the causative agent. For example, insect grazing appears to cause a larger and more widespread response in birch than does artificial damage.[47,100] Two distinct response patterns have been noted, the rapid[137] and the long-term[124] changes. Tuomi et al.[124] suggest that the long-term changes that occur in birch[53,54,100] in response to defoliation result from nutrient stress, their argument being that by removal of foliage nutrients through herbivory, the carbon/nutrient balance is altered to favor greater synthesis of phenolic compounds in remaining and subsequently developing foliage. This hypothesis may be plausible for long-term responses, but is not viable as an explanation for short-term responses.

A small number of well-publicized studies[2,52,109] have suggested that herbivory or simulated herbivory on one individual can induce synthesis of phenolic allelochemicals in other individuals not under attack. Although the large variation in concentrations of phenolic compounds occurring within and between individuals and the range of extrinsic factors which influence them make this a particularly difficult problem to study,[35] a possible mechanism for long-range induction of allelochemics does exist. Cotton leaves infected with *Aspergillus flavus* emit a volatile sesterterpene which is capable of triggering production of phenolics in the leaves of unaffected plants.[138]

G. SEASONALITY, PLANT DEVELOPMENT, PLANT HABIT, AND FORM

Even in tropical ecosystems there is seasonality, notably in rainfall, and more-temperate environments obviously undergo strong seasonal variation. Climatic variation will affect the capacity to perform photosynthesis (temperature, day length, and cloud cover) and availability of mineral nutrition (rainfall and soil temperature). Plant growth processes are closely attunded to seasonality and it is difficult to separate developmental and climatic influences on secondary metabolism.

In one of the earliest studies to take seasonality into account, Feeny and Bostock[34] found that young leaves of *Quercus robur* contained hydrolysable tannins, while condensed tannins appeared later in the growing season and then accumulated steadily. This pattern has been confirmed for other oak species.[115] In sugar maple the level of phenolics rises to a maximum in May and then remains relatively steady, but in yellow birch there is a gradual decline during the growing season.[117] Several poplar species show a decline in phenolics as leaves age,[80] and it is reported[91] that midseason gypsy moth larvae develop more rapidly on older leaves than on young. The boreal species *Betula glandulosa* and *Ledum groenlandicum* allocate considerable quantities of photosynthate to production of phenolics early in the season; synthesis of other allelochemics peaks later.[106] An examination of total and protein-binding phenolics in three North American prairie species, *Andropogon gerardii, Penstemon digitalis,* and *Lespedeza cuneata,* revealed that there were differing patterns of production through the growing season (compare the general decline with age in *A. gerardii* with summer peak in *L. cuneata).*[79] An analysis of phenol and tannin levels in boreal plants over two growing seasons confirmed that synthesis was attuned to climatic variability; in the warmer, sunnier year production was higher, presumably due to enhanced photosynthetic activity.[64] In tropical ecosystems, young leaves, particularly those produced in synchronous flushes, are often rich in phenolic compounds, notably condensed tannins.[130]

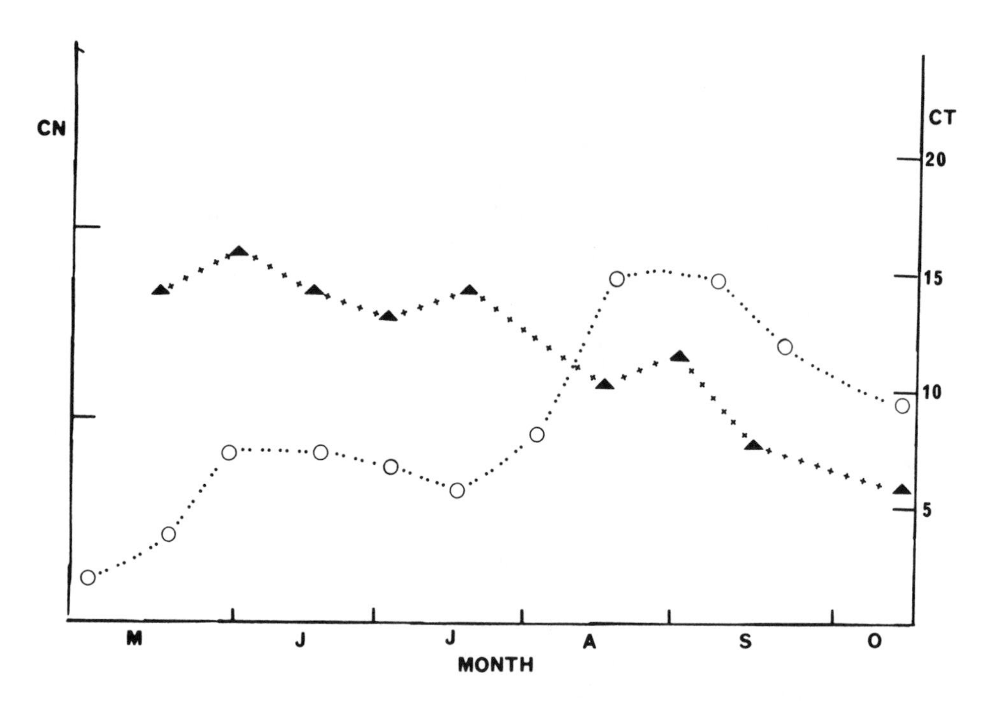

FIGURE 3. Seasonal changes in the levels of cyanide (CN, triangles) and condensed tannin (CT, circles) production in pinnules of bracken. CT is measured as percentage dry weight relative to quebracho tannin and CN as arbitrary units.

Seasonal variation is not restricted to phenolic compounds. The concentration of the alkaloid boschniakine in the leaves of *P. digitalis* declines steadily during the growing season.[79] In a study of cyanogenesis among species of the coastal restinga in Brazil,[66] the percentage of individuals reacting positive varied from 24—27% (June to July) to 53—56% (November to April). In bracken, cyanogenesis is greatest in young fronds (May), while condensed tannins become the major allelochemics late in the growing season (July to August) (Figure 3).[23] Production is reduced in shaded plants, but temporal patterns of allocation remain unchanged. The distribution of glucosinolates in *Cardamine cordifolia* shows seasonal variation between plant parts, diminishing in stems and cauline leaves over the growing season but not in basal roots or leaves.[112] Levels of both sesquiterpenes and condensed tannins are highest in leaf buds of *Hymenaea* sp. and diminish as the leaves develop.[24]

High levels of phenolics in the overwintering twigs of birch decline in the period of rapid spring growth and then rise again.[103] The inverse correlation between rapid tissue growth and production of allelochemics can be seen in vines in which levels of chlorogenic acid in shoots and stems declines in response to flower and fruit production.[78] However, it is not the case that all secondary metabolites diminish in periods of high metabolic activity; levels of alkaloids and glycosides in medicinal plants are often greatest during periods of flowering or fruiting.[122] Juvenile growth forms (root or stump sprouts) occur when some tree species are felled or damaged.[71] A number of studies suggest that such growth is often associated with a heavier investment in chemical defense than is found in the mature form.[12,16]

Karban[67] has proposed that the ability of a plant to increase resistance to herbivory may be age dependent, arguing that herbivore or pathogen attack on seedlings leads to physical, chemical, and/or physiological changes so that they become more resistant to further attack. This is proposed as a mechanism for long-term induced changes (Section II.F), but the best studies of that phenomenon[124] show a metabolic relaxation

Table 6
COMPARISON OF LEVELS OF TOTAL PHENOLICS (TP), CONDENSED TANNINS (CT), ACID DETERGENT FIBER (ADF), AND ALKALOIDS (% OF SPECIES EXAMINED) IN YOUNG AND MATURE FOLIAGE OF DECIDUOUS AND EVERGREEN SPECIES FROM SANTA ROSA NATIONAL PARK, COSTA RICA[60]

	Deciduous species		Evergreen species	
Measure	Young	Mature	Young	Mature
Number of species	49	47	12	15
TP (% dry wt)	12.9	12.4	15.8	19.0
CT (% dry wt)	10.0	10.1	9.5	13.7
ADF (% dry wt)	35.6	38.9	44.1	44.5
Alkaloids	36		14	

period during which phenolic metabolites return to preinfestation levels. It is probable that fast-growing seedlings or saplings will have to allocate their primary metabolites in a way different from the mature individual. The implications of this have not been explored for production of secondary metabolites.

Both apparency[33] and resource allocation[21] hypotheses predict differences in plant chemistry relating to plant habit; this has been confirmed by many studies. As expected, rapidly growing species (colonizers, vines, and herbaceous plants) have generally proved less liable to produce large amounts of relatively immobile quantitative defenses than climax species.[19,130] For *Cecropia peltata* grown in uniform conditions, the genetically-based variation observed for the production of condensed tannins correlated inversely with leaf production, reinforcing the idea that resources available to growth are closely related to the emphasis placed on secondary metabolism.[20]

Leaf life span can be an important factor.[12] A comparison of evergeen and deciduous species of forest trees from Costa Rica showed the former to be more likely to possess a defense chemistry based on phenolics, whereas the latter more often relies on production of alkaloids or other toxins (Table 6).[60] This dichotomy correlates well with selection of host plants by saturniid and sphingid moths, respectively. The evergreen *Ledum groenlandicum* allocates more carbon resources to synthesis of phenolic and terpenoid compounds than does the deciduous *B. glandulosa.*[106] Shaver[119] found that for the evergreen shrub *L. palustre,* NP fertilization increased leaf production but reduced leaf longevity. Unfortunately, no chemical analyses were performed, but it is anticipated that this response will be mirrored in the emphasis placed on synthesis of secondary metabolites.

As a generalization, it does appear that for leaves in particular a continuum of allelochemical investment profiles does exist in which emphasis on lignification and production of polyphenols (although not simple phenols) relates to (1) plant habit (arboreal vs. herbaceous) and (2) leaf life span (evergreen vs. deciduous). The degree to which there is a comparable emphasis on qualitative defenses (alkaloids, etc.) in herbaceous and/or deciduous plants is not yet clear, as general surveys for such compounds are more difficult to perform.

III. PATHWAYS OF SECONDARY METABOLISM

In Section II, evidence has been reviewed that establishes a range of different environmental pressures and plant or plant-part developmental processes capable of influencing the level to which constitutive allelochemicals occur in plant parts, within species, between species, and between ecosystems. The other major task of this review is to examine, at the molecular level, the biochemical processes and mechanisms that are responsible for this phenomenon. Today we know much about the biosynthetic processes that occur in plants and can predict with some certainty the routes through which most allelochemics are formed. In Section III, the major pathways and the regulatory systems that govern them are outlined.

A. GENERATION OF THE KEY INTERMEDIATES

The origin of secondary metabolites is usually considered in terms of a very limited number of precursors, typically acetyl coenzyme A (Ac-CoA), mevalonic acid (MVA), and SA. In reality the situation is more complex, but it remains one of the great marvels of the natural world that from so few building blocks, such an incredible array of end products are achieved (Scheme 1).

The starting point of plant metabolism is the incorporation of carbon dioxide into the pentose phosphate cycle. The major product exiting the cycle, glucose-6-phosphate, is subjected to the process of glycolysis which converts each hexose molecule into two units of phospho-enol pyruvate (PEP) and then into Ac-CoA. Ac-CoA is in turn incorporated into the tricarboxylic acid (TCA) cycle, but in addition it acts as the direct precursor of many secondary metabolites including a wide range of long-chain acyclic products and cyclic phenolics (polyketides). The combination of three molecules of Ac-CoA gives rise to MVA, the precursor of terpenoid secondary metabolites. The intermediates of the glycolytic pathway and of the TCA cycle are also the source of non-aromatic amino acids.

The SA pathway is rather more complex. Initially, PEP combines with erythrose-4-phosphate, a second product of the pentose phosphate cycle. This leads directly to SA. However, while SA can be the direct precursor of some important secondary metabolites such as gallic acid[50] (the phenolic precursor of hydrolysable tannins), it is itself more commonly an intermediate. Addition of a further PEP unit to SA yields chorismic acid, which is a point of bifurcation leading to either tryptophan or through phenylpyruvic acid to phenylalanine and tyrosine. It is at the chorismate or phenylpyruvate stage that the SA pathway requires an input of nitrogen — a nutrient-derived resource not involved in the Ac-CoA or MVA pathways. Nitrogen incorporated in phenylalanine is reclaimed through the enzyme phenylalanine ammonia lyase (PAL) which deaminates the amino acid to give cinnamic acid, the substrate for formation of lignin and many common phenolic secondary metabolites.

B. PROLIFERATION OF SECONDARY METABOLITES

It is not the purpose here to deal extensively with the derivation of individual secondary metabolites. There are many excellent texts dealing with metabolic pathways to secondary metabolites.[39,49,55,85] There are, however, three points that need to be made in relation to this review.

First, it is common to discuss secondary metabolites in terms of their intermediate precursors. Thus, end products are referred to as shikimate-derived, acetate-derived, mevalonate-derived, etc. It is important to remember that many end products are the result of mixed pathways (Table 7). An important example is condensed tannins, the flavan nuclei of which originate from both the shikimate pathway (60% of carbon) and the acetate pathway (40% of carbon). Changes in synthesis of such mixed metabolites require a coordinated supply of substrate from different intermediate routes.

Second, a number of features of the enzymes of secondary metabolism must be kept

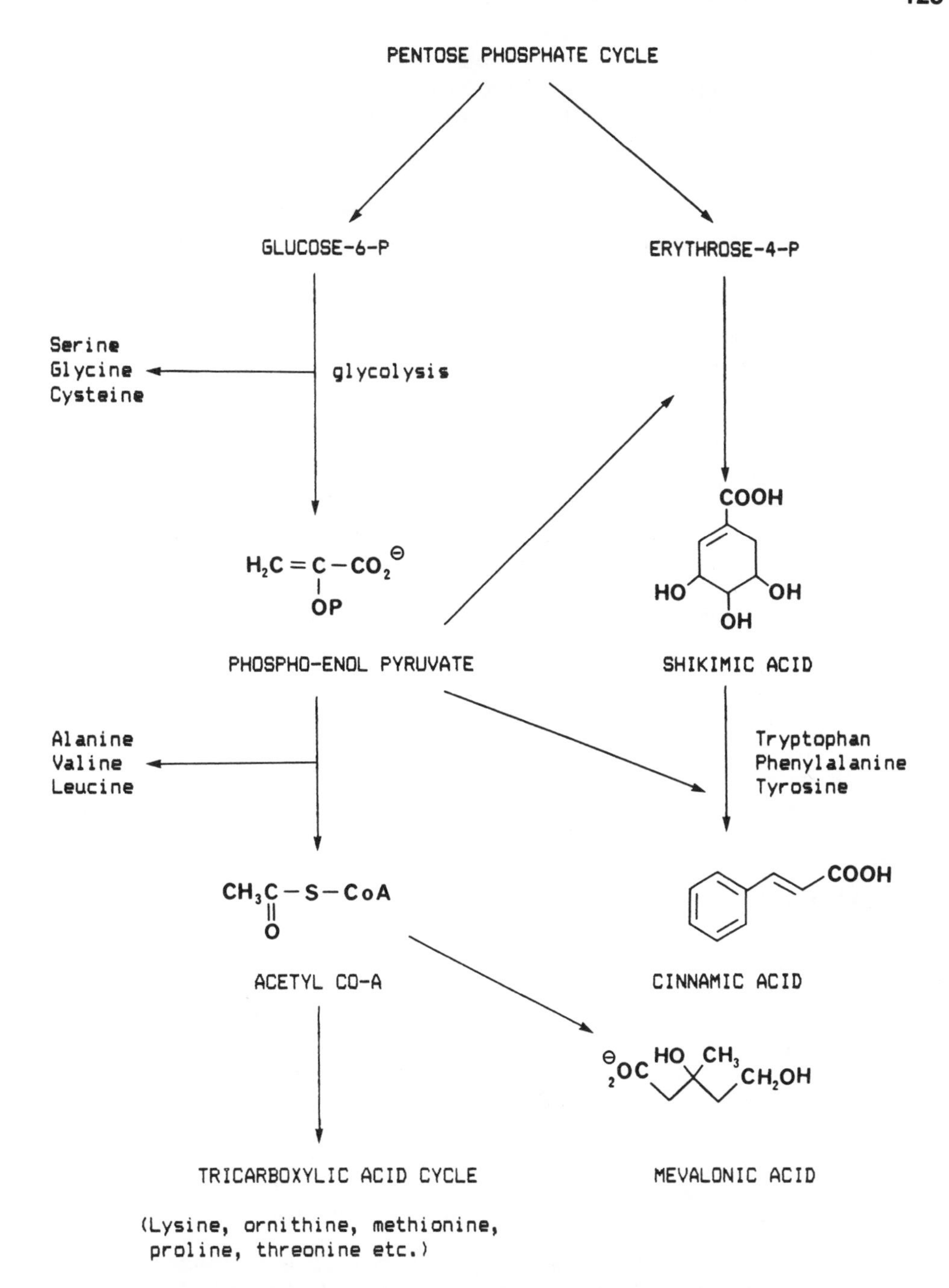

SCHEME 1. Major routes of primary and intermediate biosynthesis leading to the precursors of secondary metabolism.

in mind. The generation of secondary metabolites is always under direct genetic/enzymatic control (although final steps may be spontaneous). Specific mRNA and protein synthesis is involved in producing the enzymes concerned with the synthesis of secondary metabolites.[84] Some of these enzymes appear to be relatively nonspecific and can, with varying degrees of efficiency, catalyze a number of similar reactions.[51] Secondary metabolic pathways are often reticulate, i.e., an end product may be attained through related but different routes.[51] A good example of this is gallic acid, which can be formed either directly from SA or by β-oxidation of a phenylpropene (cinnamic acid

Table 7
SOME EXAMPLES OF GROUPS OF POTENTIAL ALLELOCHEMICS FORMED FROM MIXED BIOGENETIC PATHWAYS

	Precursor involvement			
Allelochemic type	**SA**	**Ac-CoA**	**MVA**	**AA**
Alkaloids				
Simple isoquinoline/carboline	+	+		
Indole type	+		+	
Tropane type	(+)	+		+
Flavonoids	+	+	(+)	
Condensed tannins	+	+		
Furocoumarins	+		+	
Xanthones	+	+	(+)	
Anthraquinones		+	(+)	

Note: SA = shikimic acid, Ac-CoA = acetate, MVA = mevalonic acid, and AA = nonaromatic amino acid. (+) = involved in some routes or common additions to the skeleton formed from the other precursors.

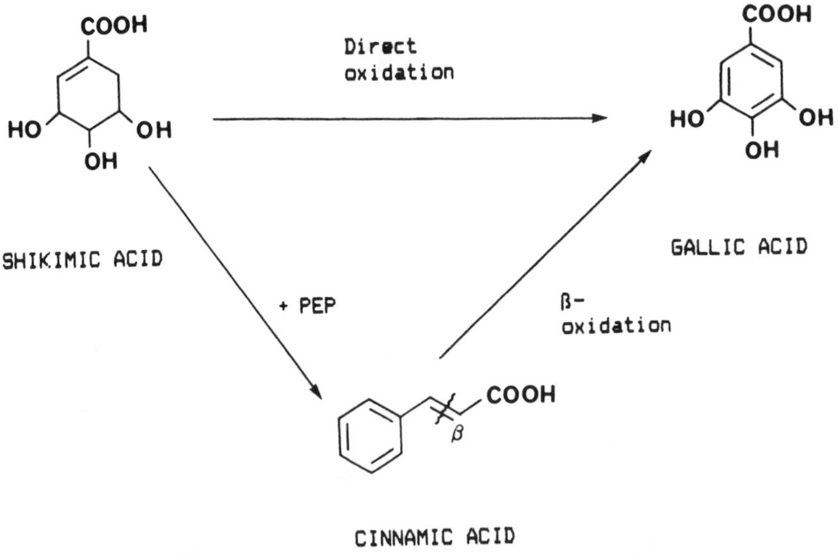

SHIKIMIC ACID
Direct oxidation
GALLIC ACID
+ PEP
β-oxidation
CINNAMIC ACID

SCHEME 2. Two routes for the production of gallic acid.

derivative). Both routes occur in *Acer buergerianum* and *Rhus succedanea* and appear to be favored at different periods of the growing season (Scheme 2).[28]

In some cases, metabolic pathways consist of coordinated associations of enzymes performing a sequence of modifications on a "bound" intermediate.[58] In such systems, a single enzyme can control the activity of a whole pathway by determining the rate of substrate supply.[58,133] Enzymes or complete enzyme systems with seemingly identical activity (isozymes) can exist in different subcellular compartments. Gottlieb[41] lists a large number of examples where enzymes of the glycolytic and pentose phosphate cycles have been found in both the plastid and cytosol. Enzymes of the SA pathway, from DAHP synthase (which unites PEP and erythrose-4-phosphate) through to those responsible for production of tryptophan, tyrosine and phenylalanine (Scheme 3), occur in both plastid and cytosol.[61,62]

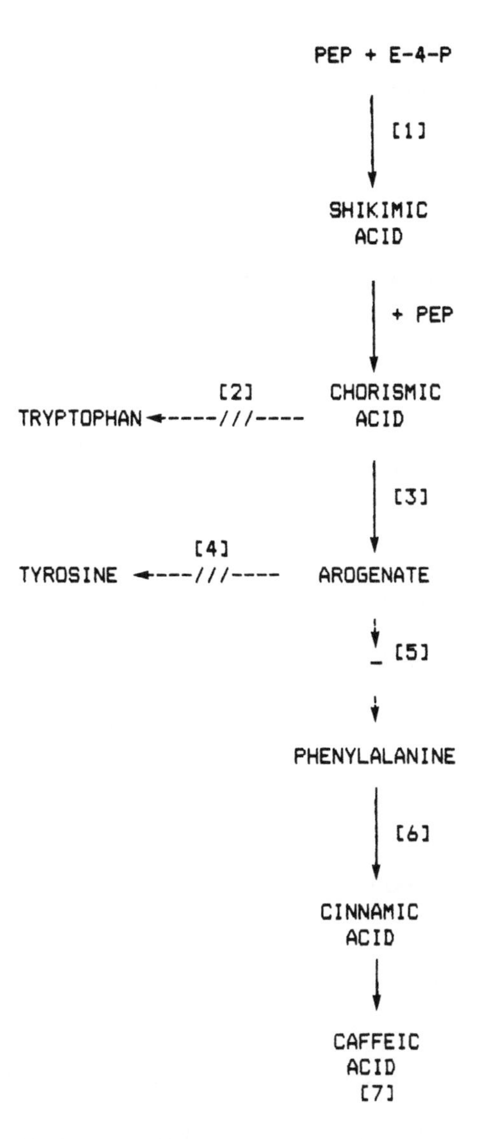

SCHEME 3. The shikimic acid pathway, some checks and balances. [1] = DAHP synthase; [2] excess tryptophan inhibits anthranilate synthase; [3] = chorismate mutase; [4] excess tyrosine inhibits arogenate dehydrogenase; [5] excess phenylalanine inhibits arogenate dehydratase; [6] = PAL; [7] excess caffeic acid can inhibit [1] and [3]. (From Jensen, R. A., *Recent Adv. Phytochem.,* 20, 57, 1986. With permission.)

Finally, the question of the "cost" of production of secondary metabolites is addressed briefly. In energy terms, their production has a clear "cost". For example, it is estimated[7] that the generation of 1 mol of tryptophan, the amino acid precursor of many alkaloids, requires 78 mol of energy-rich ATP. In terms of substrate "cost", the position is much less clear.[32] Gulmon and Mooney[44] list figures of between 2.58 and 5.00 g of CO_2 per gram as the cost of construction for a range of secondary metabolites, but it is difficult to know what to make of such values unless turnover rates and the "cost" of employing whatever is the limiting precursor can be measured. Notwithstanding the difficulty in estimating "cost", it does appear that the quantity of allelochemic produced by an individual can have an impact on growth rate[20] and that given a pest/pathogen-free environment, high levels of allelochemic synthesis may well make the producer uncompetitive. While definitive statements cannot be made, it does seem probable that substrate "costs" must be kept to a minimum by using those resources most readily available (= cheaply produced?), as presumed in resource allocation theory (Section II.C).

C. REGULATION OF BIOSYNTHETIC PATHWAYS

Despite considerable research effort, we still have a rather vague knowledge of what factors actually initiate and control secondary metabolism. It is obvious that pathways leading to the key precursors (Ac-CoA, MVA, and SA) are active at all levels of plant development as they generate products for primary functions. Among the factors thought to play a role in the regulation of secondary metabolic activity are changes in morphology, build-up in levels of intermediate substrates, and various stress factors.[51] Three separate mechanisms that are capable of regulating production are considered.

Hormonal control—Trewavas[123] has reviewed the association of different hormones with stress condition, e.g., gibberellins (light), auxin (wounding, drought, and temperature changes), abscisic acid (osmoadaptation), cytokinin (N impoverishment and shading), and ethylene (waterlogging, senescence, and wounding). Most phytohormones (cytokinin, gibberellins, absicic acid, auxin, calmodulin, and ethylene) are capable of influencing enzyme activity.[57,84,127] Lowering of exogenous auxin levels or increasing cytokinin both lead to increased PAL activity.[65] Cytokinin application can directly increase synthesis of terpenes in whole plants.[31] Light has been established as a potential initiator of enzyme activity, notably, but not exclusively, in relation to enzymes of the SA pathway, particularly PAL.[65,111,120] UV-C irradiation of leaves of *Spathiphyllum* sp. is reported[107] to cause ethylene production followed by enhanced levels of abscisic acid and indole acetic acid. The subsequent reduction in transpiration of leaves is attributed to stomatal closure initiated by simple phenolic acids whose appearance is linked with phytohormone production.

Direct control of enzyme activity— The activity of extant enzymes of primary metabolism is often governed by feedback mechanisms. That is, the catalytic activity of the enzyme is inhibited by the presence of an excess of product.[46] The same is true for some important enzymes of secondary metabolism. For example, the three amino acids produced from the SA pathway are each able to inhibit the activity of an enzyme responsible for their production (Scheme 3).[62] An important feature of this feedback mechanism is that for each amino acid, the inhibition takes place on an enzyme that is specific for its production so that a surplus of one amino acid does not inhibit the whole pathway but only that part involved in its formation. A more drastic control of enzyme activity may be achieved by a nonreversible inactivation through denaturation of the enzyme. This appears to be mediated by controlling enzymes with proteolytic activity.[128]

Control through substrate/enzyme imbalance— Another mechanism whereby formation of secondary metabolites may be controlled is through governing the accessibility of the substrate to the enzyme. For some enzymes, such as PAL, it appears that under normal physiological conditions the capacity of the extant enzyme may be substantially greater than the quantity of substrate available.[86] If this is the case, then there is a built-in capacity for immediate response to a surge in substrate levels. Da Cunha[25] has recently presented evidence that phenylpropanoid biosynthesis is regulated by the supply of amino acid and not by feedback control. It appears likely that the opposite situation can also occur— some studies of the levels of enzymes and their relation to the onset of biosynthesis suggest that there is excess substrate, so that the enzyme actually limits the rate of metabolism.[51] Substrate availability can also be regulated through compartmentation. Products produced in one compartment may be stored (possibly inhibiting further synthesis through feedback) or transferred to another compartment for further metabolism (in which case feedback inhibition at the original site would not occur). Experimental variation of membrane permeability has been shown to be able to markedly change the rate of biosynthesis.[113]

IV. EXTRINSIC FACTORS AND SECONDARY METABOLISM: TOWARDS LINKING MECHANISMS

The very abbreviated outline of secondary metabolic processes and the checks and contraints placed on the formation of allelochemicals (Section III) represents the framework from which we must understand the phenotypic and genotypic variations that have been observed (Section II). The potential mechanisms by which such changes can be brought about are clearly numerous. They include, for example, changing the rate of presentation of substrate (which may stimulate some enzymes while inhibiting others), altering the rates of enzyme formation or degradation, and changing the compartmentation of substrates within the cell. For the purposes of this discussion, a basic division is made between what are called mediated and spontaneous changes.

A. MEDIATED CHANGES

The most obvious example of mediated change is the production of phytoalexins in response to pathogen attack. This is stimulated by natural elicitors which are compounds (usually oligosaccharides) that originate from the cell wall of the pathogen or from the fragmenting cells of the host plant.[30,36,73] The mechanism of action is at the gene, leading to mRNA and protein synthesis. The products produced are novel in structure, i.e., they are not part of the constitutive secondary metabolism of the plant.

Not all elicitors lead to the synthesis of novel compounds. A glycoprotein from the rust fungus *Puccinia graminis* causes increased activity of PAL and other enzymes of the SA pathway in wheat, particularly those invovled in the biosynthesis of lignin.[94] There is an increase in extractable enzyme activity, suggesting *de novo* synthesis. Elicitor-treated cell suspensions in parsley synthesize antifungal furocoumarins (Scheme 4) derived partly from the shikimate route (via PAL) and partly from MVA;[45] in similarly treated carrot cell cultures there is also enhanced PAL activity.[82] In both cases, the effect appears to be at the level of gene transcription rather than making use of excess available enzyme capacity. Light also stimulates PAL synthesis.[65,111] In the same parsley cell suspensions noted above, light stimulation leads to the buildup of flavonoid end products rather than furocoumarins (Scheme 4).[45] Whether the end products are flavonoids or furocoumarins, PAL productivity is increased, but where light is the stimulant, the second pathway involves Ac-CoA rather than MVA. It is an interesting fact that both end products shown in Scheme 4 involve 1 mol of cinnamic acid and 3 mol of Ac-CoA (either direct or combined to give 1 mol of MVA).

It is probable that phytohormones influence metabolism by gene transcription in a manner similar to cell wall elicitors. Their effects in osmoregulation do not appear to primarily involve secondary metabolism, but their potential importance in modulating the response to wounding and nutrient stress must not be overlooked. Other areas where they might have an effect include limitation of availability of a precursor[74] by changing membrane permeability and so allowing contact between normally separated enzyme and substrate.[113] An extreme example of how compartmentation effects secondary metabolism is the case of cyanogenesis, where mechanical breakdown of cell membranes permits interaction of cyanogenic glycoside and enzyme (stored separately) to liberate HCN.[22]

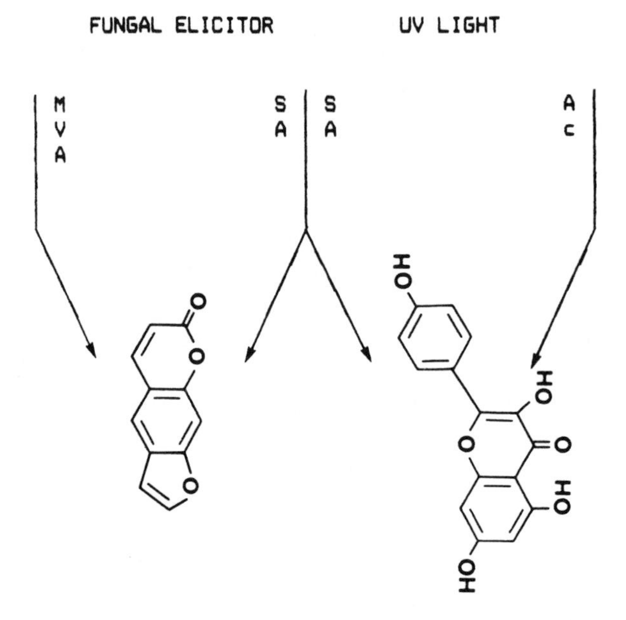

SCHEME 4. Induction of furocoumarins and flavonoids by light and fungal elicitors. MVA = mevalonic acid, SA = shikimic acid, and Ac = acetate. (From Hahlbrock, K., Chappell, J., and Kuhn, D. N., in *The Genetic Manipulation of Plants and its Application to Agriculture,* Lea, P. J. and Stewart, G. R., Eds., Clarendon Press, Oxford, 1984, 171. With permission.)

B. SPONTANEOUS CHANGES

According to Hoffmann,[57] as much as 38% of carbon fixed by photosynthesis may be lost through photorespiration. Further significant losses may occur via cyanide-resistant respiration, exudation through the root, or through entering storage pools that are not actively metabolized. Thus, it is clear that gross primary production by plants far exceeds net primary production, and the likelihood of excess carbon fixation (overflow) when environmental conditions limit growth by affecting nonphotosynthetic parts of metabolism is considerable.

The study of overflow metabolism has been largely undertaken using axenic bacterial cultures. The resulting evidence[99] is that overflow really does occur and can be rationalized simply in terms of the buildup of carbon-based (nonnitrogenous) intermediates. In nature, microbial growth is generally considered to be regulated by nutrient limitation and overflow metabolism must be a common event.[121]

The contention that overflow metabolism does occur in higher plants has been strongly argued by both Hoffmann[57] and Haslam.[49,51] If this is so, then species adapted to nutrient-deficient conditions (i.e., with metabolism leading to permanent carbon excess) can be expected to have selected secondary metabolic pathways that employ the products of overflow and to consequently have a high-profile carbon-based defense chemistry. Among species able to colonize variable environments, overflow will be a less-reliable source of substrate for secondary metabolism. Such species might exhibit the greatest degree of fluctuation in metabolic activity, the results of which could overlay the normal, less-extravagant, constitutive allelochemic profile of the plant. Species adapted to nutrient-rich environments will not normally encounter overflow conditions. In proposing that the production of some allelochemics occurs as a result of overflow, one must address the point that many plants that should do not appear to make these accumulations and thus seem not liable to overflow. The process of cyanide-resistant respiration represents an obvious nonaccumulative mechanism by which plants can divert this overflow into carbon dioxide.[57,75]

Haslam[49] has drawn attention to the fact that intermediates involved in the formation of secondary metabolites arise from the glycolytic route (Scheme 1) just prior to entry into the TCA cycle (exceptions are the amino acids lysine and ornithine). He suggests that where these intermediates build up to levels in excess of that required for primary processes, they can be diverted into the pathways of secondary metabolism. The mechanism by which this channeling takes place is unknown, but one fascinating possibility is that it invovles the cytosolic enzyme systems of glycolysis[41] and the SA pathway.[61,62] As stated by Boudet et al.[7] "It is now almost certain that the shikimate pathway is duplicated within the cell in chloroplasts and in the cytosol. . . . This separation should permit an independent regulation of each pathway and the hypothesis of their specialised role either in feeding protein or phenolic synthesis is particularly attractive."

If Boudet et al.[7] are correct, then we now have a basis for rationalizing SA overflow metabolism, and similar origins must be considered possible for Ac-CoA and MVA intermediates. While the most common route through which overflow will manifest itself should be the ubiquitous pathway leading to SA-derived simple phenolics, end products from other (including mixed) pathways will occur.

Major questions remain. For example, what is the origin of the amino acids that are widely involved in the biosynthesis of alkaloids (phenylalanine, tyrosine, and tryptophan)? Do they derive from the cytosolic pathway of SA metabolism or from the seemingly more-regulated plastid pathway? Likewise, what is the origin of the cinnamyl alcohols involved in lignification?

V. CONCLUDING REMARKS

There is substantial evidence that extrinsic factors influence levels of production of secondary metabolites and that these in turn can and do influence palatability of plant materials to herbivores. Secondary metabolism can be influenced at critical points on biosynthetic pathways by various stimulatory or antagonistic actions on enzyme activity, by variation in membrane permeability, and by changes in substrate availability.

A number of stressors influence metabolism through messenger intermediates, either hormonal or elicitor, stimulating gene transcription. The most important of these (mediated) effects for secondary metabolite production are the responses to pathogen invasion and high light intensity and perhaps the short-term induction of allelochemics in response to herbivory. However, with chewing insects, short-range induction of allelochemicals may arise simply from disruption of membranes bringing together reactants. In contrast, the major impact of nutrient stress, carbon/nutrient imbalance, and the long-term effects of defoliation should be through (spontaneous) overflow of excess fixed carbon.

REFERENCES

1. **Anderson, J. M., Proctor, J., and Vallack, H. W.,** Ecological studies in four contrasting lowland rain forests in Gunung Mulu National Park, Sarawak. III. Decomposition processes and nutrient losses from leaf litter, *J. Ecol.,* 71, 503, 1983.
2. **Baldwin, I. T. and Shultz, J. C.,** Rapid changes in leaf chemistry induced by damage: evidence for communication between plants, *Science,* 221, 277, 1983.
3. **Bate-Smith, E. C.,** Chemistry and phylogeny in the angiosperms, *Nature (London),* 269, 353, 1972.
4. **Bazzaz, F. A., Chiarello, N. R., Coley, P. D., and Pitelka, L. F.,** Allocating resources to reproduction and defense, *BioScience,* 37, 58, 1987.
5. **Bloom, A. J., Chapin, F. S., and Mooney, H. A.,** Resource limitations in plants — an economic analogy, *Annu. Rev. Ecol. Syst.,* 16, 363, 1985.
6. **Blua, M. J., Hanscom, Z., and Collier, B. D.,** Glucocapparin variability among four populations of *Isomeris arborea* Nutt, *J. Chem. Ecol.,* 14, 623, 1988.
7. **Boudet, A. M., Graziana, A., and Ranjeva, R.,** Recent advances in the regulation of the prearomatic pathway, in *The Biochemistry of Plant Phenolics,* van Sumere, C. F. and Lea, P. J., Eds., Clarenden Press, Oxford, 1985, 135.
8. **Brachet, J. and Cosson, L.,** Changes in the total alkaloid content of *Datura innoxia* Mill. subjected to salt stress, *J. Exp. Bot.,* 37, 650, 1986.
9. **Briske, D. D. and Camp, B. J.,** Water stress increases alkaloids concentration in threadleaf groundsel *(Senecio longilobus), Weed Sci.,* 30, 106, 1982.
10. **Bryant, J. P.,** Feltleaf willow-snowshoe hare interactions: plant carbon/nutrient balance and floodplain succession, *Ecology,* 68, 1319, 1987.
11. **Bryant, J. P., Chapin, F. S., and Klein, D. R.,** Carbon/nutrient balance of boreal plants in relation to vertebrate herbivory, *Oikos,* 40, 357, 1983.
12. **Bryant, J. P., Chapin, F. S., Reichardt, P. B., and Clausen, T. P.,** Adaptation to resource availability as a determinant of chemical defense strategies in woody plants, in *Chemically Mediated Interactions between Plants and Other Organisms,* Cooper-Driver, G. A., Swain, T., and Conn, E. E., Eds., Plenum Press, New York, 1985, 219.
13. **Bryant, J. P., Chapin, F. S., Reichardt, P. B., and Clausen, T. P.,** Response of winter chemical defense in Alaska paper birch and green alder to manipulation of plant carbon/nutrient balance, *Oecologia,* 72, 510, 1987.
14. **Chazdon, R. L.,** Light variation and carbon gain in rain forest understory palms, *J. Ecol.,* 74, 995, 1986.
15. **Chapin, F. S., Bloom, A. J., Field, C. B., and Waring, R. H.,** Plant responses to multiple environmental factors, *BioScience,* 37, 49, 1987.
16. **Chapin, F. S., Bryant, J. P., and Fox, J. F.,** Lack of induced chemical defense in juvenile Alaskan woody plants in response to simulated browsing, *Oecologia,* 67, 457, 1985.
17. **Chapin, F. S., van Cleve, K., and Tryon, P. R.,** Relationship of ion absorption to growth rate in taiga trees, *Oecologia,* 69, 238, 1986.
18. **Chapin, F. S., Vitousek, P. M., and van Cleve, K.,** The nature of nutrient limitation in plant communities, *Am. Nat.,* 127, 48, 1986.
19. **Coley, P. D.,** Herbivory and defensive characteristics of tree species in a lowland tropical forest, *Ecol. Monogr.,* 53, 209, 1983.
20. **Coley, P. D.,** Costs and benefits of defense by tannins in a neotropical tree, *Oecologia,* 70, 238, 1986.
21. **Coley, P. D., Bryant, J. P., and Chapin, F. S.,** Resource availability and plant antiherbivore defense, *Science,* 230, 895, 1985.
22. **Conn, E. E.,** Compartmentation of secondary compounds, in *Membranes and Compartmentation in the Regulation of Plant Functions,* Boudet, A. M., Alibert, G., Marigo, G., and Lea, P. J., Eds., Clarendon Press, Oxford, 1984, 1.
23. **Cooper-Driver, G., Finch, S., Swain, T., and Bernays, E.,** Seasonal variation in secondary plant compounds in relation to the palatability of *Pteridium aquilinum, Biochem. Syst. Ecol.,* 5, 177, 1977.
24. **Crankshaw, D. R. and Langenheim, J. H.,** Variation in terpenes and phenolics through leaf development in *Hymenaea* and its possible significance to herbivory, *Biochem. Syst. Ecol.,* 9, 115, 1981.
25. **Da Cunha, A.,** The estimation of L-phenylalanine ammonia lyase shows phenylpropanoid biosynthesis to be regulated by L-phenylalanine supply and availability, *Phytochemistry,* 26, 2723, 1987.
26. **Davies, R. I., Coulson, C. B., and Lewis, D. A.,** Polyphenols in plant, humus and soil. IV. Factors leading to an increase in biosynthesis of polyphenol in leaves and their relationship to mull and mor formation, *J. Soil Sci.,* 15, 310, 1964.
27. **del Moral, R.,** On the variability of chloro-

genic acid concentration, *Oecologia,* 9, 289, 1972.

28. **Dewick, P. M.,** The biosynthesis of shikimate metabolites, *Nat. Prod. Rep.,* 495, 1985.
29. **Dicosmo, F. and Towers, G. H. N.,** Stress and secondary metabolism in cultured plant cells, *Recent Adv. Phytochem.,* 18, 97, 1984.
30. **Ebel, J.,** Phytoalexin synthesis: the biochemical analysis of the induction process, *Annu. Rev. Phytopathol.,* 24, 235, 1986.
31. **El-Keltawi, N. E. and Croteau, R.,** Salinity depression of growth and essential oil formation in spearmint and marjoram and its reversal by foliar cytokinin, *Phytochemistry,* 26, 1333, 1987.
32. **Fagerstrom, T., Larsson, S., and Tenow, O.,** On optimal defence in plants, *Functional Ecol.,* 1, 73, 1987.
33. **Feeny, P. P.,** Plant apparency and chemical defence, *Recent Adv. Phytochem.,* 10, 1, 1976.
34. **Feeny, P. P. and Bostock, H.,** Seasonal changes in the tannin content of oak leaves, *Phytochemistry,* 7, 87, 1968.
35. **Fowler, S. V. and Lawton, J. H.,** Rapidly induced defenses and talking trees: the devil's advocate position, *Am. Nat.,* 126, 181, 1985.
36. **Friend, J.,** Phenolic substances and plant disease, in *The Biochemistry of Plant Phenolics,* van Sumere, C. F. and Lea, P. J., Eds., Clarendon Press, Oxford, 1985, 367.
37. **Frischknecht, P. M., Battig, M., and Baumann, T. W.,** Effect of drought and wounding stress on indole alkaloid formation in *Catharanthus roseus, Phytochemistry,* 26, 707, 1987.
38. **Gartlan, J. S., McKey, D. B., Waterman, P. G., Mbi, C. N., and Struhsaker, T. T.,** A comparative study of the phytochemistry of two African rain forests, *Biochem. Syst. Ecol.,* 8, 401, 1980.
39. **Geissman, T. A. and Crout, D. H. G.,** *Organic Chemistry of Secondary Plant Metabolism,* Freeman, Cooper and Co., San Francisco, 1969.
40. **Gershenzon, J.,** Changes in the levels of plant secondary metabolites under water and nutrient stress, *Recent Adv. Phytochem.,* 18, 273, 1984.
41. **Gottlieb, L. D.,** Conservation and duplication of isozymes in plants, *Science,* 216, 373, 1982.
42. **Greenway, H. and Munns, R.,** Mechanism of salt tolerance in nonhalophytes, *Annu. Rev. Plant Physiol.,* 31, 149, 1980.
43. **Gref, R. and Tenow, O.,** Resin acid variation in sun and shade needles of Scots pine (*Pinus sylvestris* L.), *Can. J. For. Res.,* 17, 346, 1987.
44. **Gulmon, S. L. and Mooney, H. A.,** Costs of defense and their effects on plant productivity, in *On the Economy of Plant Form and Function,* Givnish, T. J. and Robichaux, R., Eds., Cambridge University Press, Cambridge, 1986, 681.
45. **Hahlbrock, K., Chappell, J., and Kuhn, D. N.,** Rapid induction of mRNAs involved in defence reactions in plants, in *The Genetic Manipulation of Plants and its Application to Agriculture,* Lea, P. J. and Stewart, G. R., Eds., Clarendon Press, Oxford, 1984, 171.
46. **Hall, J. L., Flowers, T. J., and Roberts, R. M.,** *Plant Cell Structure and Metabolism,* 2nd ed., Longman, London, 1982.
47. **Hartley, S. E. and Firn, R. D.,** Phenolic biosynthesis, leaf damage and insect herbivory in birch *(Betula pendula), J. Chem. Ecol.,* in press.
48. **Hartley, S. E. and Lawton, J. H.,** Effects of different types of damage on the chemistry of birch foliage and the responses of birch feeding insects, *Oecologia,* 74, 432, 1987.
49. **Haslam, E.,** *Metabolites and Metabolism,* Clarendon Press, Oxford, 1985.
50. **Haslam, E.,** Hydroxybenzoic acids and the enigma of gallic acid, *Recent Adv. Phytochem.,* 20, 163, 1986.
51. **Haslam, E.,** Secondary metabolism — fact and fiction, *Nat. Prod. Rep.,* 217, 1986.
52. **Haukioja, E. and Hanhimaki, S.,** Rapid wound-induced resistance in white birch *(Betula pubescens)* foliage to the geometrid *Epirrita autumnata:* a comparison of trees and moths within and outside the outbreak range of the moth, *Oecologia,* 65, 223, 1985.
53. **Haukioja, E., Niemela, P., and Siren, S.,** Foliage phenols and nitrogen in relation to growth, insect damage and ability to recover after defoliation in the mountain birch *Betula pubescens* spp. *tortuosa, Oecologia,* 65, 214, 1985.
54. **Haukioja, E., Suomela, J., and Neuvonen, S.,** Long-term inducible resistance in birch foliage: triggering cues and efficacy on a defoliator, *Oecologia,* 65, 363, 1985.
55. **Herbert, R. B.,** *Biosynthesis of Natural Products,* Chapman and Hall, London, 1981.
56. **Hillis, W. E. and Swain, T.,** The phenolic constituents of *Prunus domestica.* II. The analysis of tissues of the Victoria plum tree, *J. Sci. Food Agric.,* 10, 135, 1959.
57. **Hoffmann, P.,** Ecophysiological aspects of biomass production in higher plants, *Photosynth. Res.,* 7, 3, 1985.
58. **Hrazdina, G. and Wagner, G. J.,** Compartmentation of plant phenolic compounds: sites of synthesis and accumulation, in *The Biochemistry of Plant Phenolics,* van Su-

mere, C. F. and Lea, P. J., Eds., Clarenden Press, Oxford, 1985, 119.

59. **Janzen, D. H.,** Tropical blackwater rivers, animals and mast fruiting by the Dipterocarpaceae, *Biotropica,* 6, 69, 1974.
60. **Janzen, D. H. and Waterman, P. G.,** A seasonal census of phenolics, fibre and alkaloids in foliage of forest trees in Costa Rica: some factors influencing their distribution and relation to host selection by Sphingidae and Saturniidae, *Biol. J. Linn. Soc.,* 21, 439, 1984.
61. **Jensen, R. A.,** The shikimate/arogenate pathway: link between carbohydrate metabolism and secondary metabolism, *Physiol. Plant,* 66, 164, 1985.
62. **Jensen, R. A.,** Tyrosine and phenylalanine biosynthesis: relationship between alternative pathways, regulation and subcellular location, *Recent Adv. Phytochem.,* 20, 57, 1986.
63. **Johansson, T.,** Development of stump suckers by *Betula pubescens* at different light intensities, *Scand. J. For. Res.,* 2, 77, 1987.
64. **Jonasson, S., Bryant, J. P., Chapin, F. S., and Anderson, M.,** Plant phenols and nutrients in relation to variations in climate and rodent grazing, *Am. Nat.,* 128, 394, 1986.
65. **Jones, D. H.,** Phenylalanine ammonia-lyase: regulation of its induction and its role in plant development, *Phytochemistry,* 23, 1349, 1984.
66. **Kaplan, M. A. C., Figueiredo, M. R., and Gottlieb, O. R.,** Variation in cyanogenesis in plants with season and insect pressure, *Biochem. Syst. Ecol.,* 11, 367, 1983.
67. **Karban, R.,** Herbivory dependent on plant age: a hypothesis based on acquired resistance, *Oikos,* 48, 336, 1987.
68. **Karban, R. and Carey, J. R.,** Induced resistance in cotton seedlings to mites, *Science,* 225, 53, 1984.
69. **Karunanithy, R. and Kapel, M.,** Effect of hard Lewis acids on tannin synthesis in plants — relationship of tin, bismuth and iron in *Acacia catechu, J. Pharm. Pharmac.,* 37, 44P, 1985.
70. **Kraemer, M. E., Rangappa, M., Gade, W., and Benepal, P. S.,** Induction of trypsin inhibitors in soybean leaves by Mexican bean beetle (Coleoptera: Coccinellidae) defoliation, *J. Econ. Entomol.,* 80, 237, 1987.
71. **Kramer, P. J. and Kedrowski, T. T.,** *Physiology of Woody Plants,* Academic Press, New York, 1979.
72. **Kubitzki, K. and Gottlieb, O. R.,** Phytochemical aspects of angiosperm origin and evolution, *Acta Bot. Neerl.,* 33, 457, 1984.
73. **Kuhn, D. N.,** Plant responses to stresses at the molecular level, *Plant Microb. Interact.,* 2, 414, 1986.
74. **Laloraya, M. M., Srivastan, H. N., and Guruprasad, K. N.,** Recovery of gibberellic acid inhibition of betacyanin biosynthesis by pigment precursors, *Planta,* 128, 275, 1976.
75. **Lambers, H.,** The physiological significance of cyanide resistant respiration, *Plant Cell Environ.,* 3, 293, 1980.
76. **Larsson, S., Wiren, A., Lundgren, L., and Ericsson, T.,** Effects of light and nutrient stress on leaf phenolic chemistry of *Salix dasyclados* and susceptibility to *Galerucella lineola* (Coleoptera), *Oikos,* 47, 205, 1986.
77. **Leather, S. D., Watt, A. D., and Forrest, G. I.,** Insect-induced chemical changes in young lodgepole pine *(Pinus contorta)*: the effect of previous defoliation on oviposition, growth and survival of the pine beauty moth, *Panolis flammea, Ecol. Entomol.,* 12, 275, 1987.
78. **Lilov, D. and Angelova, Y.,** Changes in the content of some phenolic compounds in connection with flower and fruit formation in vines, *Biol. Plant. (Praha),* 29, 34, 1987.
79. **Lindroth, R. L., Batzli, G. O., and Seigler, D. S.,** Patterns in the phytochemistry of three prairie plants, *Biochem. Syst. Ecol.,* 14, 597, 1986.
80. **Lindroth, R. L., Hsio, M. T. S., and Scriber, J. M.,** Seasonal patterns in the phytochemistry of three *Populus* species, *Biochem. Syst. Ecol.,* 15, 681, 1987.
81. **Lokar, L. C., Maurich, V., Mellerio, G., Moneghini, M., and Poldini, L.,** Variation in terpene composition of *Artemisia alba* in relation to environmental conditions, *Biochem. Syst. Ecol.,* 15, 327, 1987.
82. **de Lorenzo, G., Ranucci, A., Bellincampi, D., Salvi, G., and Cervone, F.,** Elicitation of phenylalanine ammonis-lyase in *Daucus carota* by oligogalacturonides released from sodium polypectate by homogenous polygalacturonase, *Plant Sci.,* 51, 147, 1987.
83. **Lowman, M. D. and Box, J. D.,** Variation in leaf toughness and phenolic content among five species of Australian rain forest trees, *Aust. J. Ecol.,* 8, 17, 1983.
84. **Luckner, M.,** Expression and control of secondary metabolism, in *Encyclopaedia of Plant Physiology,* Vol. 8, Bell, E. A. and Charlwood, B. V., Eds., Springer-Verlag, Berlin, 1980, 23.
85. **Manitto, P.,** *Biosynthesis of Natural Products,* Ellis Horwood, Chichester, 1981.
86. **Margna, U.,** Control of the level of substrate supply — an alternative in the regulation of phenylpropanoid accumulation in cells, *Phytochemistry,* 16, 419, 1977.
87. **Mattson, W. J. and Haack, R. A.,** The role of drought in outbreaks of plant-eating insects, *BioScience,* 37, 110, 1987.
88. **McDonald, A. J. S., Ericsson, A., and Lohammar, T.,** Dependence of starch storage

on nutrient availability and photon flux density in small birch (*Betula pendula* Roth.), *Plant Cell Environ.*, 9, 433, 1986.

89. **McKey, D. B.,** The distribution of secondary compounds within plants, in *Herbivores: Their Interactions with Plant Secondary Compounds,* Rosenthal, G. A. and Janzen, D. H., Eds., Academic Press, New York, 1979, 56.
90. **Medina, E., Olivares, E., and Diaz, M.,** Water stress and light intensity effects on growth and nocturnal acid accumulation in a terrestrial CAM bromeliad (*Bromelia humilis* Jacq.) under conditions, *Oecologia,* 70, 441, 1986.
91. **Meyer, G. A. and Montgomery, M. E.,** Relationships between leaf age and food quality of cottonwood foliage for the gypsy moth, *Lymantria dispar, Oecologia,* 72, 527, 1987.
92. **Mihaliak, C. A., Couvet, D., and Lincoln, D. E.,** Inhibition of feeding by a generalist insect due to increased volatile leaf terpenes under nitrate-limiting conditions, *J. Chem. Ecol.,* 13, 2059, 1987.
93. **Mihaliak, C. A. and Lincoln, D. E.,** Growth pattern and carbon allocation to volatile leaf terpenes under nitrogen-limiting conditions in *Heterotheca subaxillaris* (Asteraceae), *Oecologia,* 66, 423, 1985.
94. **Moerschbacher, B., Heck, B., Kogel, K. H., Obst, O., and Reisener, H. J.,** An elicitor of the hypersensitive lignification response in wheat leaves isolated from the rust fungus *Puccinia graminis* f. sp. *tritici.* II. Induction of enzymes correlated with the biosynthesis of lignin, *Z. Naturforsch.,* 41C, 839, 1986.
95. **Mole, S., Ross, J. A. M., and Waterman, P. G.,** Light-induced variation in phenolic levels in foliage of rain forest plants. I. Chemical changes, *J. Chem. Ecol.,* 14, 1, 1988.
96. **Mole, S. and Waterman, P. G.,** Light-induced variation in phenolic levels in foliage of rain forest plants. II. Potential significance to herbivores, *J. Chem. Ecol.,* 14, 23, 1988.
97. **Myers, J. H. and Williams, K. S.,** Lack of short or long term inducible defenses in the red alder — western tent caterpillar system, *Oikos,* 48, 75, 1987.
98. **Nascimento, J. C. and Langenheim, J. H.,** Leaf sesquiterpene and phenolics in *Copaifera multijuga* on contrasting soil types in a central Amazonian rain forest, *Biochem. Syst. Ecol.,* 14, 615, 1986.
99. **Neijssel, O. M. and Tempest, D. W.,** The physiology of metabolite over-production, *Symp. Soc. Gen. Microbiol.,* 29, 53, 1977.
100. **Neuvonen, S., Haukioja, E., and Molarius, A.,** Delayed inducible resistance against a leaf-chewing insect in four deciduous tree species, *Oecologia,* 74, 363, 1987.
101. **Newbery, D. McC. and de Foresta, H.,** Herbivory and defense in pioneer, gap and understory trees of tropical rain forest in French Guiana, *Biotropica,* 17, 238, 1985.
102. **Paine, T. D. and Stephen, F. M.,** Influence of tree stress and site quality on the induced defense system of loblolly pine, *Can. J. For. Res.,* 17, 569, 1987.
103. **Palo, R. T., Sunnerheim, K., and Theander, O.,** Seasonal variation of phenols, crude protein and cell wall content in birch (*Betula pendula* Roth.) in relation to ruminant *in vitro* digestibility, *Oecologia,* 65, 314, 1985.
104. **Pearcy, R. W., Bjorkman, O., Caldwell, M. M., Keeley, J. E., Monson, R. K., and Strain, B. R.,** Carbon gain by plants in natural environments, *BioScience,* 37, 21, 1987.
105. **Pizzi, A. and Cameron, F. A.,** Flavonoid tannins — structural wood components for drought-resistance mechanisms of plants, *Wood Sci. Technol.,* 20, 119, 1986.
106. **Prudhomme, T. I.,** Carbon allocation to antiherbivore compounds in a deciduous and an evergreen subarctic shrub species, *Oikos,* 40, 344, 1983.
107. **Rajagopal, R., Ulvskov, P., Marcussen, J., Andersen, J. M., and Allerup, S.,** Hormonal and phenolic changes accompanying and following UV-C induced stress in *Spathiphyllum* leaves, *J. Plant Physiol.,* 130, 291, 1987.
108. **Rajaratnam, J. A. and Hook, L. I.,** Effect of boron nutrition on intensity of red spider mite attack on oil palm seedlings, *Exp. Agric.,* 11, 59, 1975.
109. **Rhoades, D. F.,** Pheromonal communication between plants, *Recent Adv. Phytochem.,* 19, 195, 1985.
110. **Rhoades, D. F. and Cates, R. G.,** Towards a general theory of plant antiherbivore chemistry, *Recent Adv. Phytochem.,* 10, 168, 1976.
111. **Rhoades, M. J. C.,** The physiological significance of plant phenolic compounds, in *The Biochemistry of Plant Phenolics,* van Sumere, C. F. and Lea, P. J., Eds., Clarenden Press, Oxford, 1985, 99.
112. **Rodman, J. E. and Louda, S. M.,** Seasonal flux of isothiocyanate-yielding glucosinolates in roots, stems and leaves of *Cardamine cordifolia, Biochem. Syst. Ecol.,* 13, 405, 1985.
113. **Roos, W., Furst, W., and Luckner, M.,** Significance of membrane permeability for the synthesis and distribution of alkaloids in cultures of *Penicillium cyclopium* Westling, *Nova Acta Leopold.,* 7, 175, 1976.
114. **Ryan, C. A., Bishop, P. D., Graham, J. S.,**

Broadway, R. M., and Duffey, S. S., Plant and fungal cell wall fragments activate expression of proteinase inhibitor genes for plant defense, *J. Chem. Ecol.,* 12, 1025, 1986.

115. **Scalbert, A. and Haslam, E.,** Polyphenols and chemical defense in the leaves of *Quercus robur, Phytochemistry,* 26, 3191, 1987.
116. **Schultz, J. C. and Baldwin, I. T.,** Oak leaf quality declines in response to defoliation by gypsy moth larvae, *Science,* 217, 149, 1982.
117. **Schultz, J. C., Nothnagle, P. T., and Baldwin, I. T.,** Seasonal and individual variation of leaf quality of two northern hardwood tree species, *Am. J. Bot.,* 69, 753, 1982.
118. **Seemann, J. R., Sharkey, T. D., Wang, J.-L., and Osmond, C. B.,** Environmental effects on photosynthesis, nitrogen-use efficiency, and metabolite pools in leaves of sun and shade plants, *Plant Physiol.,* 84, 796, 1987.
119. **Shaver, G. R.,** Mineral nutrition and leaf longevity in the evergreen shrub, *Ledum palustre* spp. *palustre, Oecologia,* 49, 362, 1981.
120. **Smith, H., Billett, E. E., and Giles, A. B.,** The photocontrol of gene expression in higher plants, in *Regulation of Enzyme Synthesis and Activity in Higher Plants,* Smith, H., Ed., Academic Press, London, 1977, 93.
121. **Tempest, D. W., Neijssel, O. M., and Zevenboom, W.,** Properties and performance of micro-organisms in laboratory cultures: their relevance to growth in natural ecosystems, *Symp. Soc. Gen. Microbiol.,* 34, 119, 1983.
122. **Trease, G. E. and Evans, W. C.,** *Pharmacognosy,* 11th ed., Balliere Tindall, London, 1978, 176.
123. **Trewavas, A.,** Resource allocation under poor growth conditions. A major role for growth substances in developmental plasticity, *Symp. Soc. Exp. Biol.,* 40, 31, 1986.
124. **Tuomi, J., Niemela, P., Haukioja, E., Siren, S., and Neuvonen, S.,** Nutrient stress: an explanation for plant anti-herbivore resposnes to defoliation, *Oecologia,* 61, 208, 1984.
125. **Uegaki, R., Kubo, S., and Fujimori, T.,** Stress compounds in the leaves of *Nicotiana undulata* induced by TMV inoculation, *Phytochemistry,* 27, 365, 1988.
126. **Valentine, H. T., Wallner, W. E., and Wargo, P. M.,** Nutritional changes in host foliage during and after defoliation and their relation to the weight of gypsy moth pupae, *Oecologia,* 57, 298, 1983.
127. **Varner, J. E. and Ho, D. T.-H.,** Hormonal control of enzyme activity in higher plants, in *Regulation of Enzyme Synthesis and Activity in Higher Plants,* Smith, H., Ed., Academic Press, London, 1977, 83.
128. **Wallace, W.,** Proteolytic inactivation of enzymes, in *Regulation of Enzyme Synthesis and Activity in Higher Plants,* Smith, H., Ed., Academic Press, London, 1977, 177.
129. **Walters, T. and Stafford, H. A.,** Variability in accumulation of proanthocyanidins (condensed tannins) in needles of douglas-fir *(Pseudotsuga menziesii)* following long-term budworm defoliation, *J. Chem. Ecol.,* 10, 1469, 1984.

129a. **Waterman, P. G.,** unpublished data.

130. **Waterman, P. G. and McKey, D. B.,** Secondary compounds in rain-forest plants: patterns of distribution and ecological implications, in *Ecosystems of the World, Tropical Rain Forest,* Leith, H. and Werger, O. R., Eds., Elsevier, Amsterdam, in press.
131. **Waterman, P. G. and Mole, S.,** Soil nutrients and plant secondary compounds, in *Mineral Nutrients in Tropical Forest and Savanna Ecosystems,* Proctor, J., Ed., Blackwell Scientific, Oxford, in press.
132. **Waterman, P. G., Ross, J. A. M., and McKey, D. B.,** Factors affecting levels of some phenolic compounds, digestibility and nitrogen content of the mature leaves of *Barteria fistulosa* (Passifloraceae), *J. Chem. Ecol.,* 10, 387, 1985.
133. **Welch, G. R. and Gaertner, F. H.,** Coordinate activation of multienzyme complex by the first substrate, *Arch. Biochem. Biophys.,* 172, 476, 1976.
134. **Whitney, G. C.,** A demographic analysis of the leaves of open and shade grown *Pinus strobus* L. and *Tsuga canadensis* (L.) Carr, *New Phytol.,* 90, 447, 1982.
135. **Wolfson, J. L.,** Developmental responses of *Pieris rapae* and *Spodoptera eridania* to environmentally induced variation in *Brassica nigra, Environ. Entomol.,* 11, 207, 1982.
136. **Woodhead, S.,** Environmental and biotic factors affecting the phenolic content of different cultivars of *Sorghum bicolor, J. Chem. Ecol.,* 7, 1035, 1981.
137. **Wratten, S. D., Edwards, P. J., and Dunn, I.,** Wound-induced changes in the palatability of *Betula pubescens* and *Betula pendula, Oecologia,* 61, 372, 1984.
138. **Zeringue, H. J.,** Changes in cotton leaf chemistry induced by volatile elicitors, *Phytochemistry,* 26, 1357, 1987.

5 Arthropod Impact on Plant Gas Exchange

Stephen C. Welter
Department of Entomology
University of California, Berkeley
Berkeley, California

TABLE OF CONTENTS

I. INTRODUCTION

Numerous reviews of arthropod impact on plants within native and agricultural systems have been undertaken. Interactions between deleterious herbivore effects and changes in plant structure, chemistry, or phenology have been discussed in a number of reviews or articles.[16,53,73,78,90] Similarly, papers have addressed potential beneficial aspects of herbivory[66-69] or plant compensating mechanisms to minimize herbivore impact.[25,59-61] Belsky[7] provides a critical evaluation of the exisitng evidence supporting beneficial aspects of herbivory. Most reviews have dealt primarily with questions of arthropod impact on plant fitness (as implied by changes in growth or reproductive output) or crop losses. However, several reviews have also briefly considered physiological consequences of herbivory.[24,41,60]

Understanding basic physiological responses to herbivory may provide a key to both understanding and predicting herbivore impact. Examination of gas exchange processes provides a common biological basis for comparison of herbivore impact on different plants or plant parts. If researchers are not going to be fettered forever to repeating experiments for each plant-arthropod combination, then a more general understanding of plant responses is required. Impact of arthropods on photosynthesis has already been incorporated into numerous plant models as part of simulation modeling efforts such as those of Coughenour,[15] van Roermund et al.,[96] or Gutierrez et al.[34] This review encompasses only the effects of arthropods on plant photosynthesis, respiration, and water-vapor exchange.

Herbivory has been categorized within this chapter by feeding-damage type as suggested by Root et al.[80] rather than taxonomic grouping. Defoliation studies, defined as complete or partial direct leaf removal, include both natural defoliation and clipping experiments. Additional groupings include mesophyll feeders, gall formers, epidermal feeders, phloem feeders, stem borers, and root feeders. Arthropods that feed on specific structures within a plant such as the phloem are collectively termed selective tissue feeders in contrast to defoliators. This is not to imply that defoliators cannot exhibit selectivity in terms of leaf or site selection.

The advantage of a guild approach may be illustrated easily by examination of the family of leafhoppers, Cicadellidae. Individuals within the subfamily Typhlocybinae feed exclusively on leaf mesophyll cells, individuals within the subfamily Cicadellinae feed on the xylem, and members of the subfamily Deltocephalinae feed on the phloem.[94] Therefore, leafhoppers within the subfamily Typhlocybinae are discussed within this review in conjunction with feeding damage by spider mites and leaf miners, whereas the subfamilies Deltacephalini would be included with other phloem feeders such as aphids.

Literature for this review has been gleaned from both managed systems and ecological studies on native plants. The extent to which agricultural and native plant literature sources are pooled within various reviews appears to depend on the particular interests or bias of the authors rather than concrete differences between systems. One argument against pooling data sets has been the potentially legitimate fear that plants under intense agricultural breeding programs no longer are true models for prediction of native systems.[7] However, this apprehension must not blind researchers to the possibility that common processes do exist between systems and that discovery of differences between systems may lay the groundwork for understanding individual systems. Studies are required that directly compare native and agricultural systems such that areas inappropriate for data pooling may be identified and avoided.

II. REVIEW OF TERMINOLOGY

The two major processes reviewed in this chapter are photosynthesis and transpiration, whereas discussion of respiratory processes is rather limited due to limited available information. For a review of terminology and theory associated with gas exchange processes, see Larcher[51] or Farquhar and Sharkey.[26]

Photosynthesis involves enzymatic, photochemical, and diffusion processes. Each of these processes may be limited by either internal or external factors. Herbivore impact is discussed in reference to these limiting processes.

Simplistically, gas flux from a leaf may be expressed by the following equation:

$$J = \frac{\Delta C}{\Sigma r} \tag{1}$$

where J represents gas flux from a leaf and ΔC is the concentration difference between outside air and at the site of the reaction. Σr represents the sum of the terms representing resistance to diffusion. These resistance terms include boundary layer resistance (r_a), cuticular resistance (r_c), stomatal resistant (r_s), intercellular resistance (r_i), and mesophyll resistance (r_m). Boundary layer resistance involves the resistance to gas diffusion due to higher gas concentration within the thin layer of air immediately surrounding the leaf. The thickness of the layer depends on a variety of factors including leaf pubescence, air movement, leaf size, and leaf position. Cuticular resistance is associated with the resistance to gas exchange provided by the leaf cuticule. Cuticular resistance in daylight is generally quite high for most situations. Stomatal resistance is defined as the limitation to gas exchange due to stomatal closure. Intercellular resistance depends on the specific intercellular volume and pathways available for gas diffusion. Mesophyll resistance is a residual term that includes CO_2 diffusion to the site, resistance at the gas-fluid interface, resistance to carboxylation, and other biochemical processes. Transpiration from a leaf is restricted by cuticular resistance, boundary layer resistance, intercellular resistance, and stomatal resistance. Photosynthesis is limited by all resistance values. The inverse of resistance terms are termed conductance values. Therefore, an increase in a resistance value has a negative impact on either water vapor exchange or CO_2 uptake.

III. IMPACT OF FEEDING GUILDS

A. DEFOLIATORS

Nonselective tissue removal of either partial or entire leaves by arthropods with grinding mandibular surfaces are included under defoliation studies. Selective tissue feeding by mandibulate arthropods such as lepidopterous leaf miners is not included within this category. Studies involving plant tissue clipping to stimulate insect damage or larger herbivore grazing are included. However, clipping studies may not reflect some arthropod damage accurately.[74]

1. Increases in CO_2 Exchange Rates

Arthropod feeding damage by complete tissue removal appears to generally cause an increase in photosynthetic rates per unit area (Figure 1). The increase in photosynthesis will generally be less than 50%,[22,88] but as great as two- to threefold increase.[83] Causal mechanisms have been divided into environmental modifications (extrinsic factors) and physiological responses (intrinsic mechanisms).[60] Extrinsic mechanisms have

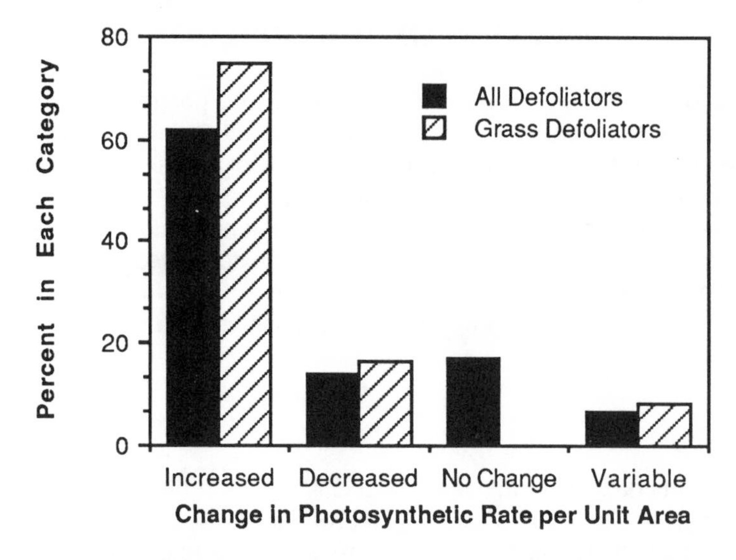

FIGURE 1. Direction of change in photosynthetic rates per unit area in remaining leaf tissue after defoliation.

included increased light penetration after defoliation; improved plant water status, improved plant nutritional status; and direct effects of hormones, hormone analogs, or growth factors from herbivore secretions.

Several different intrinsic mechanisms have been proposed. Neales and Incoll[63] reviewed one hypothesis that photosynthetic potential may be limited in a negative feedback system by accumulation of assimilates at the source site. Increases in sink demands or decreases in other assimilate sources by defoliation will place greater demands for assimilates on remaining leaf tissue. An increase in assimilate demand per leaf after defoliation and the subsequent increased translocation of assimilates from remaining leaf tissue prevent the inhibition of photosynthesis due to end-product accumulation. Increases in photosynthesis due to increased demand for assimilates have been tested in several ways. Increased demand for assimilates from unshaded source leaves was generated by shading of adjacent leaves, stems and pods.[93] The increase in assimilate demand resulted in decreased starch concentration, increased sucrose concentrations, and increased activity of the carboxylating enzyme ribulose-1,5-diphosphate in unshaded leaves. Partial defoliation in several species led to increased assimilate demand on remaining foliage and increased photosynthesis per unit leaf area or per unit weight.[11,22,33,100] Decreases in sink demand by elimination of growing apices led to a reduction in photosynthetic rate per unit dry leaf weight. Reductions of adjacent assimilate sources by leaf removal led to increases in photosynthetic rates in *Pinus radiata.*[91] Partially defoliated bean plants showed increases in photosynthesis and ribulose-1,5-diphosphate carboxylase activity in remaining leaf tissue, but no increase in stomatal resistance to CO_2 diffusion.[100] Similarly, increases in assimilate demand resulted from increased regrowth following defoliation. New tissue is also younger and more photosynthetically active than the previously defoliated tissue. Thus, apparent increases in photosynthesis may also result from changes in mean age of photosynthetically active tissue.[11,20,28,70]

Increases in enzyme synthesis in remaining foliage on partially defoliated plants were hypothesized to result from reduced competition between leaves for mineral nutrients of cytokinin, a plant hormone.[100] Cytokinins are growth regulators that have been shown to cause increased stomatal opening. Increased cytokinin levels and photosynthetic rates have been reported after artificial defoliation.[22,89,106]

Delays in senescence of remaining leaf tissue after defoliation have also been widely reported.[11,33,42,64,65,89,106] Older leaf tissue has also been shown to be rejuvenated to rates similar to those of young, fully expanded leaves.[42,43] These increases have been as great as two- to threefold. The increases in photosynthetic rates of older remaining leaves or delayed senescence have been associated with either decreases in mesophyll resistance[33,43,89] or stomatal resistance.[20,42] In addition, increased chlorophyll production per unit area has been reported, but since photosynthesis and chlorophyll covaried, no change in CO_2 assimilated per unit chlorophyll occurred.[89]

The previous studies on gas exchange impact were carried out over less than 1 year, but long-term effects have also been reported. Prudhomme[76] indicated with correlative data that previous defoliation history impacted on photosynthetic rates 1 year later. Birch trees defoliated the previous year by the winter moth exhibited significantly higher photosynthetic rates than undamaged trees. Given the correlative nature of the study, the results should be interpreted as a potential response rather than a proven conclusion.

Alternatively, rather than internal physiological changes, external changes in the plant environment have been proposed for increased photosynthetic rates per unit leaf area. Increased light penetration due to defoliation has been reported.[2,47] Other hypotheses that have been proposed include enhanced water status of remaining tissue due to decreased transpirational surfaces,[61] direct effects of growth regulators produced within saliva, and improved nutrient availability.[13] These hypotheses have been less well documented.

2. Decreases or No Change in CO_2 Exchange Rate

Studies have also shown decreases in photosynthetic rates per unit area (Figure 1).[1,21,27,35,52,74] These decreases were shown to be only temporary in several papers. Similarly, papers have also shown no change in photosynthesis per unit area.[18,92]

Methodology appears to play an important role in structuring the conclusions and inferences derived by the various authors. Whereas photosynthesis per unit area has been shown to increase on many occasions, the total canopy exchange rate, if measured, has consistently been shown to decrease. Studies such as that by Detling et al.[22] showed an increase of 21% per unit area after defoliation, but an overall decrease by 20% in total canopy assimilation. Other studies have also shown decreases in total canopy production despite an increase in photosynthetic rates of the remaining leaf tissue.[8,45] Reductions in total canopy assimilation were caused by decreased leaf area indexes, smaller leaf size, and/or decreased light interception.[8,45,72,82,88] One of the few studies to show net increases in total canopy photosynthesis after clipping was by Alexander and McCloud.[2]

Therefore, the question that the authors are interested in should determine the type of measurements taken or interpretation of the data. If the question is to examine the effects of herbivory on the productivity or "fitness" of a plant, then total canopy measurements would provide a better indicator. If the authors are interested in specific plant buffering mechanisms, then use of individual leaves should provide a more detailed understanding. Given that changes in allocation to roots, reserves, or fruiting often covaries with defoliation, an increase in photosynthetic rates per unit area does not necessarily translate into an increased fitness. However, the increase in photosynthesis per unit area may act to partially ameliorate the deleterious effects of herbivory.

An increase in photosynthesis should not necessarily be construed as an evolved response to herbivory. Any increases in photosynthesis following defoliation may be the result of the plant returning to a balance between supply and demand of its various tissues. Plants appear to maintain a degree of homeostasis between assimilate supply and demand. Therefore, herbivory may be just one of many factors that alter this balance.

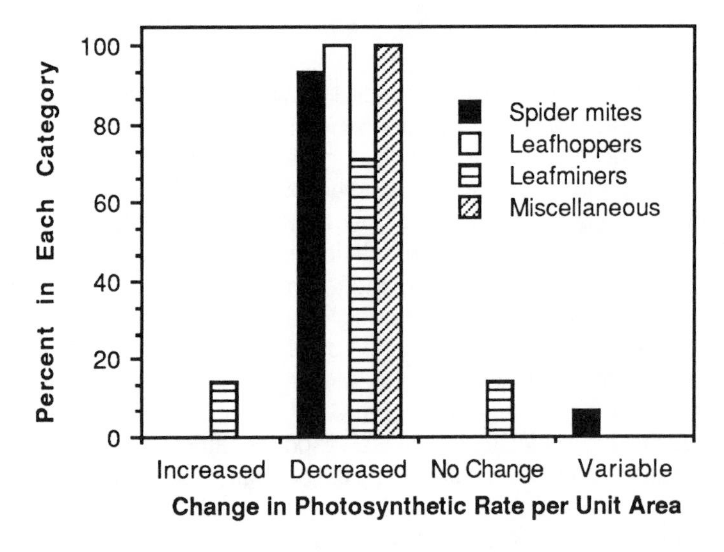

FIGURE 2. Direction of change in photosynthetic rates per unit area for leaf tissue damaged by various mesophyll feeding groups.

This is evidenced by studies that have used alternative methods such as shading to alter the balance between plant source and sinks.

3. Other Gas Exchange Processes

Changes in the various respiratory processes in response to herbivory have proven variable. Photorespiration in lucerne leaves following defoliation was either increased[42] or not affected.[22] Dark respiration in apples was decreased initially, then rose to higher levels than the control by day 7.[35] Similarly, soybean and barley respiration rates were depressed after defoliation.[45,82] Studies by Detling et al.[21] did not show any change in dark respiration. Root respiration was decreased in *Dactylis glomerata* due to hypothesized limited substrate availability.[18] Detling et al.[21] showed a decrease in root respiration rates.

Changes in water gas exchange have been less well studied. Increases in water loss per leaf have been observed.[37,92] No change in water use efficiency was observed in one study,[65] whereas another study showed a 20% decrease in water use efficiency of citrus leaves.[92] No predictable pattern can be discerned with so few data currently available.

B. MESOPHYLL FEEDERS

Arthropods that feed on the mesophyll and palisade layers of leaves include a wide variety of taxonomic groups including Acari, Diptera, Homoptera, Hemiptera, Lepidoptera, and Hymenoptera as some of the more important examples. Given that these taxa, as well as other feeding types, feed on a particular type of leaf tissue, these herbivores are referred to as selective tissue feeders. The majority of the data concerning these more selective feeders were developed within agricultural systems. The degree to which agricultural data may be useful for interpreting herbivore impact in native systems is poorly understood.

The impact of spider mites appears less variable than defoliators (Figure 2). The data for spider mite impact is derived exclusively from agricultural sources. Spider mites are most commonly thought of as secondary pest upsets due to predator elimination by pesticides. Thus, spider mite outbreaks are less common in native systems and apparently less well studied. Spider mites pierce the leaf epidermis with needlelike mouthparts and feed primarily on mesophyll and palisade layers of the leaf.[83]

With few exceptions,[46] spider mite feeding damage has been shown to significantly reduce host photosynthetic rates. Reductions in photosynthesis by spider mites has been shown in a wide variety of crops including peppermint, strawberries, almonds, cotton, and peaches.[5,10,19,58,62,85,86] However, the degree of reduction and the shape of the relationships have not been consistent. Early studies by Boulanger[9] using excised leaves demonstrated significant reductions in apple leaf photosynthesis and transpiration by European red mite damage, *Panonychus ulmi* Koch. Significant negative, linear correlations between total number of mites and apple photosynthesis were also obtained.[36] Reductions of up to 43% were observed at high levels of mite damage. No interaction was detected between leaf age and photosynthetic losses by European red mite.

Studies on impact of the two-spotted mite, *Tetranychus urticae* Koch, on strawberries demonstrated both linear and curvilinear reductions in photosynthesis.[83-85] The reduction in photosynthesis was shown to result from decreased stomatal opening and increased mesophyll resistance. Increases in mesophyll resistance were attributed, in part, to decreases in chlorophyll content per leaf area.[83] Kolodziej et al.[49] also examined *T. urticae* impact on strawberries and demonstrated reduced photosynthesis, reduced photorespiration, and no significant change in dark respiration. Photorespiration appears to be less sensitive to damage than photosynthesis. Similar results were obtained for almonds, peaches, and avocados.[5,58,62,86,109] Differential impact of various mite genera on almond leaf gas exchange was demonstrated.[109] Both Andrews and LaPre[5] on almonds and Mizell et al.[62] on peaches showed curvilinear reductions in leaf gas exchange.

Spider mite effects on leaf transpiration rates generally show decreased water loss rates associated with spider mite damage.[5,9,62,84,85,87] However, other studies have also shown increased nighttime transpiration rates.[19] DeAngelis et al.[19] showed decreased daytime transpiration due to decreased stomatal conductance to water, but increased water loss across the cuticular boundary at night. However, the data are presented as a percent increase in conductance rather than absolute values. Given the normal excessively low levels of cuticular conductance, the 350% increase in cuticular conductance may not translate into biologically significant water vapor exchange. Atanasov[6] showed a threefold increase in transpiration after feeding damage by *T. turkestani (atlanticus)*. However, no statistical analyses are presented with the manuscript. Thus, for spider mite impact, most papers seem to agree that photosynthesis is reduced, but impact on transpiration may vary. Too few papers deal with spider mite impact on respiratory processes to draw general conclusions.

Various abiotic interactions between spider mite feeding damage and gas exchange rates have also been examined. Significant interactions between feeding damage and temperature have been shown for cotton.[10] Spider mite impact on stomatal and mesophyll resistance values was shown to be more detrimental at 20 and 25°C than at 30°C. Photosynthesis seemed to be equally depressed at all temperatures. No data were presented to show the potential impact of temperature on mite feeding behavior or rate of feeding. No significant interactions between water stress and spider mite damage on apple leaves were observed.[27] Conversely, significantly greater reductions in photosynthesis were associated with spider mite damage on water-stressed almond trees within 1 of 2 years of study.[108]

Interactions between leaf nitrogen levels, feeding by pecan leaf scorch mite, *Eotetranychus hicoriae* (McGregor), and pecan leaf photosynthesis were demonstrated. No effects of pecan leaf scorch mite feeding damage were measurable at low nitrogen levels, whereas pecan leaf photosynthetic rates were higher for mite-damaged leaves than control leaves at the highest nitrogen level. At optimal leaf nitrogen levels, photosynthesis was significantly reduced by mite feeding damage. However, the interpretation of the study is clouded by the differential levels of mite feeding damage due to different mite population growth rates under different nitrogen regimes. In addition, only two replicates of each treatment were monitored; thus, statistical conclusions are at best limited.

Other arthropods that feed predominantly on mesophyll tissue have also been examined in their effects on gas exchange (Figure 2). Reductions of 40% in photosynthesis of Scotch pine seedlings infested by a diaspidid scale, the pine needle scale, *Phenacaspis pinifoliae* (Fitch), were measured.[99] The reductions were assumed to result from reductions in chlorophyll content in the fascicles, whereas reduced light interception and stomatal closure were considered less-like causal agents. Another diaspidid scale, the euonymus scale, *Unaspis euonymi* (Comstock), reduced the transpiration rate of its host by increasing stomatal resistance.[14] Changes in stomatal resistance for infested euonymus leaves exhibited a 2-h lag behind noninfested leaves. The stomatal resistance remained consistently higher throughout the day. The higher resistance values were attributed to possible impairment of guard cell functioning by scale feeding damage.[14] The increase in stomatal resistance resulted in decreased transpiration rates and increased leaf temperature of infested leaves. Leafhopper species have been shown to decrease photosynthetic rates of various crops including apple,[56] alfalfa,[103] and potato.[50,98] Transpiration rates decreased as leafhopper damage increased due to increasing total resistance (boundary and stomatal diffusion resistance).[103]

Leaf-mining dipteran species also have demonstrated significant influences on host gas exchange rates. Studies have demonstrated disproportionate losses in photosynthesis resulting from small losses in leaf area by *Liriomyza* spp.[48,95] Both mesophyll and stomatal conductance in adjacent leaf tissue were reduced by leaf miner feeding. Parrella et al.[71] showed that tissue mining by *L. trifolii* in the upper palisade layers caused significantly greater reductions in photosynthesis, stomatal conductance, and mesophyll conductance than *L. huidobrensis* that mines the lower spongy mesophyll layers. *Liriomyza* feeding damage on lima beans caused minor photosynthetic losses in damaged leaves ($<10\%$), but these losses were compensated for by increased photosynthetic rates in undamaged leaves.[57] The increase in photosynthetic rates was not the result of increased activity of carboxylase. No significant impact on total canopy photosynthesis for alfalfa resulted from damage by *Agromyza frontella* (Rondani).[17] Several hypotheses were presented to expalin how *A. frontella* could exploit its host without significant losses in photosynthesis. Leaves damaged by the leaf miner were generally within the shaded layers of the leaf canopy by the time the larval mines reached maximum size. Photosynthetic compensation mechanisms within alfalfa were also proposed.

Lepidopterous leaf miners also have been shown to reduce photosynthetic rates of their hosts. Photosynthesis was reduced in apple leaves as the result of increased mesophyll resistance. The increase in mesophyll resistance was attributed to increased levels of internal CO_2 and reduced chlorophyll levels.[75] No effects on stomatal conductance or transpiration were observed. Significant negative curvilinear correlations were obtained between the pear leaf miner, *Bucculatrix pyrivorella* Kuroko, and pear photosynthesis.[29]

C. GALL FORMERS

Gall formation on grapes by phylloxera, *Dactylosphaera vitifolii* Shimer, caused dramatic reductions in photosynthesis of up to 97.5% in the gall tissue.[79] As shown with some other herbivores, the galls proved to be strong sinks for assimilates from surrounding leaf tissue. Similarly, *Phylloxera notabilis* increased the stomatal and mesophyll resistance of pecan foliage.[4] The increase in resistance values was attributed to decreases in leaf chlorophyll content.

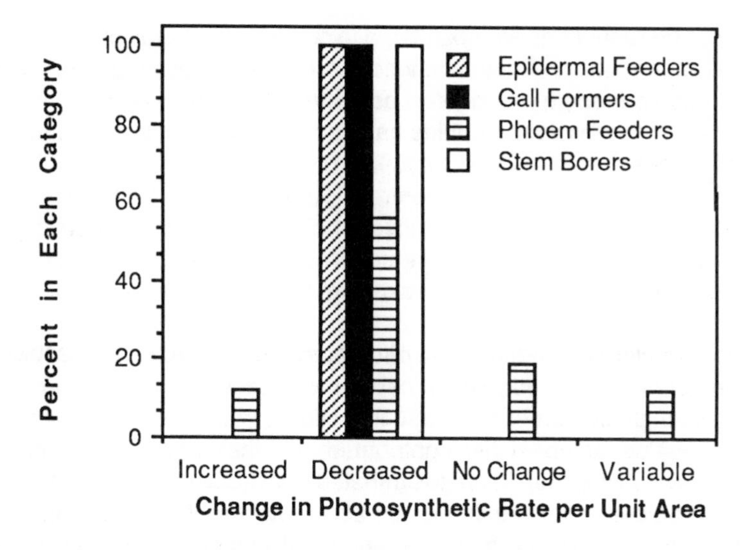

FIGURE 3. Direction of change in photosynthetic rates per unit area for leaf tissue damaged by various selective tissue feeding groups.

D. EPIDERMAL FEEDERS

Feeding by the citrus rust mite, an eriophyiid *Phyllocoptruta oleivora* (Ashmead), was shown to increase water loss for attached citrus fruit.[3] The citrus rust mite feeds on the epidermal cell layer of citrus fruit and causes a discoloration known as sharkskin, russet, or bronzing. Given that eriophyiids have shorter stylet lengths (10 μm) than spider mites (100 μm), the eriophyiid is not capable of feeding on the photosynthetically active mesophyll and palisade layers. The impact of these mites on leaf photosynthetic rates is not known. In contrast to the eriophyiid mites that possess needlelike mouthparts, other epidermal feeders such as thrips feed by rasping away the entire epidermal layer with rasping-sucking mouthparts. In a nonmanipulative study, Castro[12] correlated increasing thrips damage with decreased photosynthesis of peanuts. The palm thrips, *Thrips palmi,* has also been shown to reduce the photosynthetic rate and stomatal conductance of cucumber plants.[102] Highly significant negative correlations were obtained between percent leaf area scarred and photosynthesis or stomatal conductance. Impact of epidermal feeders on photosynthesis is shown in Figure 3. These patterns reflect only two studies.

E. PHLOEM FEEDERS

Phloem feeding arthropods have proven slightly more variable in their impact on plants than mesophyll feeders (Figure 3). In response to aphid feeding, photosynthetic rates per unit area have been shown to increase,[101,77] to not change,[81,104,105] or to decrease.[31,81,97,104,105] What is perhaps the most detailed study on aphid impact on leaf gas exchange demonstrated reductions in photosynthetic rates, transpiration rates, stomatal conductance, and total chlorophyll for cultivars of wheat susceptible to the greenbug, *Schizaphis graminum* (Rondani).[81] No changes in water use efficiency, internal CO_2 levels, or specific photosynthetic activity of chlorophyll were observed. Given the lack of change in internal CO_2 levels, the reduction in stomatal conductance does not appear to be increasing stomatal limitations to photosynthesis. However, resistant cultivars only showed reductions in transpiration, stomatal conductance, and carboxylation efficiency. The saturation level of CO_2 for photosynthesis in aphid-damaged leaves occurred at approximately two thirds the level of CO_2 for undamaged leaves. The lower saturation point indicates a reduction in photosynthetic capacity associated with mesophyll tissue. The reductions are postulated to occur as the result of decreased ribulose bisphosphate regeneration. Photosynthetic reductions associated with chlorophyll reductions from aphid feeding damage on potatoes were also reported.[32]

The excretion of honeydew by aphids has had variable effects on leaf gas exchange of wheat leaf blades. Presence of honeydew from aphids did not appear to significantly influence stomatal conductance.[81] In contrast, Rabbinge et al.[77] showed that manually applied applications of honeydew reduced wheat leaf blade photosynthetic rates by 25%, whereas 28% of the stomata were covered. Honeydew was also correlated with enhanced leaf aging and chlorophyll breakdown. However, these results were not found in the field studies.

Significant reductions in pecan leaf photosynthesis by black pecan aphid feeding damage resulted from clogged phloem cells as evidenced by direct monitoring of photosynthesis and examination of histological sections.[104] Interveinal damage by salivary toxins of the black pecan aphid also contributed to further reductions in photosynthesis. The reductions in translocation of photosynthates from actively photosynthesizing leaves due to clogged phloem cells resulted in an accumulation of photosynthates in undamaged leaves. Similar results were obtained on potato for feeding damage by the potato aphid, *Macrosiphum euphorbiae* (Thomas).[97] Radiolabeled assimilates were shown to be translocated more slowly from aphid-damaged leaves, thus resulting in higher carbohydrate levels in the leaves. The higher levels of carbohydrates within the adjacent leaf tissues were assumed to cause end-product inhibition of photosynthesis.

Aphid infestations were also shown to act as additional sinks for assimilates from adjacent leaves.[101] Assimilate transport to aphid-damaged leaves was increased by 70- to 80-fold, whereas assimilation rates were increased by approximately 6-fold in older leaves. Rabbinge et al.[77] also showed a 12.4% increase in photosynthesis. Depending on the leaf monitored, photosynthesis was depressed on leaves with infestations of *Aphis craccivora,* whereas adjacent uninfested leaves showed an increase of 18 to 34% over control leaves on uninfested plants.[107]

Aphid feeding was not shown to increase water uptake of potato plants.[30] However, interaction of greenbug damage and water stress of wheat appears to overwhelm the homeostatic capacity of the plant such that additional reductions in net assimilation rates, total chlorophyll content, internal CO_2 levels, and specific photosynthetic activity of chlorophyll occur.[81]

Whereas Wood et al.[104] showed a 25% reduction in leaf respiration rates, Hawkins et al.[38] showed an increase in total shoot (leaf, stem and petiole) respiration rate. A significant reduction was shown in total root respiration, presumably due to reduced availability of translocates from the above-ground portions of the plant.[39]

Another phloem feeder, a psyllid, *Strophingia ericae,* showed nonsignificant decreases in photosynthesis and increases in leaf respiration rates.[44]

F. STEM BORERS

Infestation by the dogwood borer, *Thamnosphecia scitula,* into dogwood trees caused increases in stomatal resistance greater than twofold, significant reductions in photosynthesis, and no change in dark respiration.[40] These changes were hypothesized to result from the disruption of the vascular tissue such that changes occur in mineral and metabolite distribution, hastened leaf senescence, and stomatal closure. However, caution must be excerised when generalizing from these data, given that only one true replicate was used for each treatment. Damage by the wood wasp, *Sirex noctilo,* caused impairment of translocation, increased transpiration and phloem respiration, and increased water loss through the bark.[54] The pattern of stem-borer impact on photosynthesis is compared in Figure 3 to other selective tissue feeders.

Table 1
CHANGES IN PHOTOSYNTHESIS PER UNIT AREA FOR DIFFERENT HERBIVORE FEEDING TYPES

Herbivore feeding type	Total number	Photosynthesis per unit area (number of papers [%]) Increased	Decreased	No change	Variable response[a]
Mesophyll feeders (all)	27	1 (3.7)	24 (88.9)	1 (3.7)	1 (3.7)
Spider mites	15	0 (0.0)	14 (93.3)	0 (0.0)	1 (6.7)
Leafhoppers	4	0 (0.0)	4 (100.0)	0 (0.0)	0 (0.0)
Leaf miners	7	1 (14.3)	5 (71.4)	1 (14.3)	0 (0.0)
Miscellaneous	1	0 (0.0)	1 (100.0)	0 (0.0)	0 (0.0)
Epidermal feeders	2	0 (0.0)	2 (100.0)	0 (0.0)	0 (0.0)
Gall formers	2	0 (0.0)	2 (100.0)	0 (0.0)	0 (0.0)
Phloem feeders	16	2 (12.5)	9 (56.3)	3 (18.8)	2 (12.5)
Stem borers	1	0 (0.0)	1 (100.0)	0 (0.0)	0 (0.0)
Defoliators (all)	29	18 (62.1)	4 (13.8)	5 (17.2)	2 (6.9)
Defoliators (grass)	12	9 (75.0)	2 (16.7)	0 (0.0)	1 (8.3)
Total	77	21 (27.3)	42 (53.2)	9 (13.0)	5 (6.5)

[a] Indicates the response to herbivory varied due to another parameter such as cultivar, leaf age, or time of sampling.

G. ROOT FEEDERS

Generally, herbivory below ground has been less well documented, presumably due to logistical problems associated with root feeders. Below-ground simulated damage caused reductions in net photosynthesis in *Boutela gracilis* as well as shifts in higher carbon allocation to roots after day 3.[23]

IV. CONCLUSIONS

The effects of defoliation appear to be quite different from those caused by selective tissue feeder damage. Feeding by mesophyll feeders including spider mites, leaf miners, leafhoppers, and other miscellaneous taxa has consistently reduced photosynthetic rates per unit area. Of the 27 studies reviewed, 88.9% showed a significant reduction (Table 1). The only exceptions were Marten and Trumble[57] and Daley and McNeil.[17] Jackson and Hunter[46] showed varied responses depending on nitrogen fertilization regimes.

Of the 29 papers on defoliation effects, 62.1% showed an increase in the photosynthetic rate per unit area, while only 13.8% showed decreases in photosynthesis per unit area. If only grass-herbivore interactions are considered, then 75% of all papers showed an increase in photosynthetic rates per unit area after defoliation, while only 16.7% showed a decrease. Phloem feeders such as aphids have proven more variable. Of all phloem feeding studies, 56.3% showed a decrease in photosynthetic rates, 12.5% showed an increase, 18.8% showed no change and 12.5% were varied in their response. If herbivore feeding type is ignored, then 53.2% of all papers reported decreases, 27.3% indicated increases, 13% showed no change, and 6.5% showed variable responses.

One question that should come from these data is "Why should selective tissue feeders not commonly cause the apparent increases in photosynthesis per unit area associated with defoliators?" Why don't the arguments of limited competition for cytokinins or increased demand for assimilates apply to selective tissue feeders?

Selective tissue feeding does not generally open up the plant canopy for increased light penetration. Several additional explanations involve potential methodological differences between defoliation and selective tissue damage studies. These differences are as follows:

1. Studies on selective tissue feeding generally monitor leaves that have been directly damaged by herbivore feeding (e.g., stippled, mined, or rasped).
2. Studies on selective tissue feeding have often had both the damaged and control treatments on the same plant.
3. Studies on selective tissue feeding have averaged data from both damaged and undamaged leaves on arthropod-infested plants, thus masking any potential increases in adjacent undamaged leaves.

With few exceptions (such as Marten and Trumble[57] or Wu and Thrower[107]), studies examining the effects of selective tissue feeding have not looked at the effects of feeding damage on undamaged leaves on a damaged plant relative to undamaged leaves on a completely undamaged plant. Two basic approaches have been used for manipulating selective tissue feeder damage. One type of study has manipulated damage levels on individual leaves and compared gas exchange rates against an undamaged leaf on the same plant. When individual leaves on the same plant are used for the control, the researcher does not have any reference against which to measure relative changes. If undamaged portions of the plant do increase photosynthetic rates, then effects of selective arthropod damage would appear exaggerated. The other class of studies has manipulated damage levels for an entire plant and determined a mean gas exchange rate for the entire plant. These approaches would also tend to mask any increased or compensatory responses by the host plant. Studies that have taken average gas exchange rates for an entire plant would fail to discern any potential increase in the remaining undamaged leaf tissue.

Defoliation studies generally compare the effects on damaged and undamaged plants. Therefore, there is an inherent bias towards comparing remaining leaf tissue (clipped or nonclipped) against undamaged leaf tissue on a nondefoliated plant.

Additional differences between selective tissue feeders and defoliators may include less impact by selective tissue feeders on extrinsic factors such as light interception. Unless infestations are severe, foliage damaged by selective tissue feeders will remain on the plant and not open up the canopy increased light penetration. Similarly, stimulation of new growth after damage by selective tissue feeders is less well documented. Thus, shifts in mean leaf age towards new growth following damage might not be expected for selective tissue feeders.

Perhaps one additional problem has been the lack of information flow between ecological and agricultural literature pools. The literature concerning selective tissue feeders comes predominantly from entomologists working in agriculture, while information on defoliators comes from both ecological and agricultural sources. Whereas data from selective tissue feeding may not be immediately applicable to a particular defoliation study, attempts to generalize about herbivore impact on the ecophysiology of native systems may be biased if data from agricultural studies on selective tissue feeding damage are not considered. Furthermore, if generalizations about the physiological consequences of different types of feeding are to be made, direct experimental comparisons of herbivore feeding types must be completed.

Future studies should investigate relationships between herbivory effects and potentially moderating factors such as nutrient levels, water status, or interplant competition. Experimental variables should span the range observed in both native and agricultural systems such that differences between these systems might be better understood.

Moreover, the validity of comparing data between these sytems should be assessed through controlled comparisons of herbivore impact on native plants and their agricultural counterparts.

REFERENCES

1. **Alderfer, R. G. and Eagles, C. F.,** The effect of partial defoliation on the growth and photosynthetic efficiency of bean leaves, *Bot. Gaz. (Chicago),* 137, 351, 1976.
2. **Alexander, C. W. and McCloud, D. E.,** CO_2 uptake (net photosynthesis) as influenced by light intensity of isolated Bermuda grass leaves contrasted to that of swards under various clipping regimes, *Crop Sci.,* 2, 132, 1962.
3. **Allen, J. C.,** The effect of citrus rust mite damage on citrus fruit drop, *J. Econ. Entomol.,* 71, 746, 1978.
4. **Andersen, P. C. and Mizell, R. F., III,** Physiological effects of galls induced by *Phylloxera notabilis* (Homoptera Phylloxeridae) on pecan foliage, *Environ. Entomol.,* 16, 264, 1987.
5. **Andrews, K. L. and LaPre, L. F.,** Effects of Pacific spider mite on physiological processes of almond foliage, *J. Econ. Entomol.,* 72, 651, 1979.
6. **Atanasov, N.,** Physiological functions of plants as affected by damage caused by *Tetranychus atlanticus,* in Proc. 3rd Int. Congr. Acarology, 1973, 183.
7. **Belsky, A. J.,** Does herbivory benefit plants? A review of the evidence, *Am. Nat.,* 127, 870, 1986.
8. **Boote, K. J., Jones, J. W., Smerage, G. H., Barfield, C. S., and Berger, R. D.,** Photosynthesis of peanut canopies as affected by leafspot and artificial defoliation, *Agron. J.,* 72, 247, 1980.
9. **Boulanger, L. W.,** The effect of European red mite feeding injury on certain metabolic activities of red delicious apple leaves, *Maine Agric. Exp. Stn. Bull.,* 570, 5, 1956.
10. **Brito, R. M., Stern, V. M., and Sances, F. V.,** Physiological response of cotton plants to feeding of three, *Tetranychus* spider mite species (Acari: Tetranychidae), *J. Econ. Entomol.,* 79, 1217, 1986.
11. **Caldwell, M. M., Richards, J. H., Johnson, D. A., and Dzurec, R. S.,** Coping with herbivory: photosynthetic capacity and resource allocation in two semiarid *Agropyron* bunchgrass, *Oecologia,* 50, 14, 1981.
12. **Castro, P. R. C., Pitelli, R. A., and Passilongo, R. L.,** Incidence of some peanut pests as affected by the stage of growth of the plant, *An. Soc. Entomol. Bras.,* 1, 5, 1972.
13. **Chew, R. M.,** Consumers as regulators of ecosystems: an alternative to energetics, *Ohio J. Sci.,* 74, 359, 1974.
14. **Cockfield, S. D. and Potter, D. A.,** Interaction of euonymus scale (Homoptera Diaspididae) feeding damage and severe water stress on leaf abscission and growth of *Euonymus fortunei, Oecologia,* 71, 41, 1986.
15. **Coughenour, M. B.,** A mechanistic simulation analysis of water use leaf angles and grazing in East African graminoids, *Ecol. Model,* 26, 203, 1984.
16. **Crawley, M. J.,** Hervivory: the dynamics of animal-plant interactions, in *Studies in Ecology,* Vol. 10, Anderson, D. J., Greig-Smith, P., and Pitelka, F. A., Eds., University of California Press, Berkeley, 1983, 437.
17. **Daley, P. F. and McNeil, J. N.,** Canopy photosynthesis and dry matter partitioning of alfalfa infested by the alfalfa blotch leafminer (*Agromyza frontella* (Rondani)), *Can. J. Plant Sci.,* 67, 433, 1987.
18. **Davidson, J. L. and Milthorpe, F. L.,** The Effect of defoliation on the carbon balance in *Dactylis glomerata, Ann. Bot. (London),* 30, 185, 1966.
19. **DeAngelis, J. D., Larson, K. C., Berry, R. E., and Krantz, G. W.,** Effects of spider mite injury on transpiration and leaf water status in peppermint, *Environ. Entomol.,* 11, 975, 1982.
20. **Deinum, B.,** Photosynthesis and sink size: an explanation for the low productivity of grass swards in autumn, *Neth. J. Agric. Sci.,* 24, 238, 1976.
21. **Detling, J. K., Dyer, M. I., and Winn, D. T.,** Effect of simulated grasshopper grazing on carbon dioxide exchange rates of western wheatgrass leaves, *J. Econ. Entomol.,* 72, 403, 1979.
22. **Detling, J. K., Dyer, M. I., and Winn, D. T.,** Net photosynthesis root respiration and regrowth of *Bouteloua gracilis* following simulated grazing, *Oecologia,* 41, 127, 1979.
23. **Detling, J. K., Winn, D. T., Procter-Gregg, C., and Painter, E. L.,** Effects of simulated grazing by belowground herbivores on growth, carbon dioxide exchange and carbon allocation patterns of *Bouteloua gracilis, J. Appl. Ecol.,* 17, 771, 1980.
24. **Dirzo, R.,** Insect-plant interactions some ecophysiological consequences of herbivory, *Illus. ISBN,* 1984.
25. **Dyer, M. I., Detling, J. K., Coleman, D. C.,**

and Hilbert, D. W., The role of herbivores in grasslands, *Grasses and Grasslands Systematics and Ecology,* Estes, J. R., Tyrl, R. J., and Brunken, J. N., Eds., University of Oklahoma Press, Norman, 1982, 255.
26. **Farquhar, G. D. and Sharkey, T. D.,** Stomatal conductance and photosynthesis, *Annu. Rev. Plant Physiol.,* 33, 317, 1982.
27. **Ferree, D. C. and Hall, F. R.,** Effects of soil water stress and twospotted spider mites on net photosynthesis and transpiration of apple leaves, *Photosynth. Res.,* 1, 189, 1980.
28. **Fuess, F. W. and Tesar, M. B.,** Photosynthetic efficiency, yields, and leaf loss in alfalfa, *Crop Sci.,* 8, 159, 1968.
29. **Fujiie, A.,** Ecological studies on the population of the pear leaf miner, *Bucculatrix pyrivorella,* (Lepidoptera Lyonetiidae). VI. Effects of injury by the pear leaf miner on leaf fall and photosynthesis of the pear tree, *Appl. Entomol. Zool.,* 17, 188, 1982.
30. **Galecka, B.,** Effect of aphid feeding on the water uptake by plants and on their biomass, *Ekol. Pol.,* 25, 531, 1977.
31. **Gerloff, E. D. and Ortman, E. E.,** Physiological changes in barley induced by greenbug feeding stress, *Crop Sci.,* 11, 174, 1971.
32. **Gibson, R. W., Whitehead, D., Austin, D. J., and Simkins, J.,** Prevention of potato top-roll by aphicide and its effect on leaf area and photosynthesis, *Ann. Appl. Biol.,* 82, 151, 1976.
33. **Gifford, R. M. and Marshall, C.,** Photosynthesis and assimilate distribution in *Lolium multiflorum* Lam. following differential tiller defoliation, *Aust. J. Biol. Sci.,* 26, 517, 1973.
34. **Gutierrez, A.,** Energy allocation in plants and animals, *Can. Entomol.,* xx, xx, 1987.
35. **Hall, F. R. and Ferree, D. C.,** Effects of insect injury simulation on photosynthesis of apple leaves, *J. Econ. Entomol.,* 69, 245, 1976.
36. **Hall, F. R. and Ferree, D. C.,** Influence of two-spotted spider mite populations on photosynthesis of apple leaves, *J. Econ. Entomol.,* 68, 517, 1975.
37. **Hammond, R. B. and Pedigo, L. P.,** Effects of artificial and insect defoliation on water loss excised soybean leaves, *J. Kans. Entomol. Soc.,* 54, 331, 1981.
38. **Hawkins, C. D. B., Aston, M. J., and Whitecross, M. I.,** Short-term effects of infestation by two aphid species on plant growth and shoot respiration of three legumes, *Physiol. Plant.,* 68, 329, 1986.
39. **Hawkins, C. D. B., Aston, M. J., and Whitecross, M. I.,** Short-term effects of two aphid species on plant growth and root respiration of three leguminous species, *Physiol. Plant.,* 67, 447, 1986.
40. **Heichel, G. H. and Turner, N. C.,** Physiological responses of dogwood, *Cornus florida,* to infestation by the dogwood borer, *Thamnosphecia scitula, Ann. Appl. Biol.,* 75, 401, 1973.
41. **Hewett, E. W.,** Some effects of infestation on plants: a physiological viewpoint, *N.Z. Entomol.,* 6, 235, 1977.
42. **Hodgkinson, K. C.,** Influence of partial defoliation on photosynthesis, photorespiration and transpiration by lucerne leaves of different ages, *Aust. J. Plant Physiol.,* 1, 561, 1974.
43. **Hodgkinson, K. C., Smith, N. G., and Miles, G. E.,** The photosynthetic capacity of stubble leaves and their contribution to growth of the lucerne plant after high level cutting, *Aust. J. Agric. Res.,* 23, 225, 1972.
44. **Hodkinson, I. D.,** The population dynamics and host plant interactions of *Strophingia ericae* (Curt.) (Homoptera psylloidea), *J. Anim. Ecol.,* 42, 565, 1973.
45. **Ingram, K. T., Herzog, D. C., Boote, K. J., Jones, J. W., and Barfield, C. S.,** Effects of defoliating pests on soybean canopy CO_2 exchange and reproductive growth, *Crop Sci.,* 21, 961, 1981.
46. **Jackson, P. R. and Hunter, P. E.,** Effects of nitrogen fertilizer level on development and populations of the pecan leaf scorch mite (Acari Tetranychidae), *J. Econ. Entomol.,* 76, 432, 1983.
47. **Jameson, D. A.,** Responses of individual plants to harvesting, *Bot. Rev.,* 29, 532, 1963.
48. **Johnson, M. W., Welter, S. C., Toscano, N. C., Ting, I. P., and Trumble, J. T.,** Reduction of tomato leaflet photosynthesis rates by mining activity of *Liriomyza sativae* (Diptera Agromyzidae), *J. Econ. Entomol.,* 76, 1061, 1983.
49. **Kolodziej, A., Kropczynska, D., and Poskuta, J.,** Comparative study on carbon dioxide exchange rates of strawberry and chrysanthemum plants infested with *Tetranychus urticae,* Akademiai Kiado: Budapest, Hungary, *Illus. ISBN,* 1979.
50. **Ladd, T. L., Jr. and Rawlins, W. A.,** The effects of the feeding of the potato leafhopper on photosynthesis and respiration in the potato plant, *J. Econ. Entomol.,* 58, 623, 1965.
51. **Larcher, W.,** *Physiological Plant Ecology,* 2nd ed., Springer-Verlag, New York, 1980, chap. 3 and 5.
52. **Li, J. R. and Proctor, J. T. A.,** Simulated pest injury effects photosynthesis and transpiration of apple leaves, *Hort-Science,* 19, 815, 1984.
53. **Louda, S. M.,** Herbivore effect on stature, fruiting, and leaf dynamics of a native crucifer, *Ecology,* 65, 1379, 1984.
54. **Madden, J. L.,** Physiological reactions of *Pinus radiata* to attack by wood wasp *Sirex noctilio* (Hymenoptera Siricidae), *Bull. Entomol. Res.,* 67, 405, 1977.
55. **Mallott, P. G. and Davy, A. J.,** Analysis of

effects of the bird cherry-oat aphid on the growth of barley unrestricted infestation, *New Phytol.*, 80, 209, 1978.

56. **Marshall, E. G. and Childers, C.,** The effects of leafhopper feeding injury on apparent photosynthesis and transpiration of apple leaves, *J. Agric. Res.*, 65, 265, 1942.
57. **Marten, B. and Trumble, J. T.,** Structural and photosynthetic compensation for leafminer (Diptera Agromyzidae) injury in lima beans, *Environ. Entomol.*, 16, 374, 1987.
58. **McClernan, W. A. and Marini, R. P.,** Effect of European red mite on photosynthesis chlorophyll content and specific leaf weight of peach leaves, *HortScience*, 20, 188, 1985.
59. **McNaughton, S. J.,** Compensatory plant growth as a response to herbivory, *Oikos*, 40, 329, 1983.
60. **McNaughton, S. J.,** Grazing as an optimization process: grass-ungulate relationships in the Serengeti, *Am. Nat.*, 113, 691, 1979.
61. **McNaughton, S. J.,** Physiological and ecological implications of herbivory, *Encycl. Plant Physiol.*, new ser. 12C, 657, 1983.
62. **Mizell, R. R., Anderseon, P. C., and Schiffhauer, D. E.,** Impact of the twospotted spider mite on some physiological processes of peach, *J. Agric. Entomol.*, 3, 143, 1986.
63. **Neales, T. F. and Incoll, L. D.,** The control of leaf photosynthesis rate by the level of assimilate concentration in the leaf: a review of the hypothesis, *Bot. Rev.*, 34, 107, 1968.
64. **Neales, T. F., Treharne, K. J., and Wareing, P. F.,** A relationship between net photosynthesis, diffusive resistance, and carboxylating enzyme activity in bean leaves, *Photosynth. Photorespir.*, 89, 1971.
65. **Nowak, R. S. and Caldwell, M. M.,** A test of compensatory photosynthesis in the field: implications for herbivore tolerance, *Oecologia*, 61, 311, 1984.
66. **Owen, D. F. and Weigert, R. G.,** Beating the walnut tree: more grass/grazer mutualism, *Oikos*, 39, 115, 1982.
67. **Owen, D. F. and Wiegart, R. G.,** Do consumers maximize plant fitness?, *Oikos*, 27, 488, 1976.
68. **Owen, D. F. and Weigert, R. G.,** Grasses and grazers: is there mutualism?, *Oikos*, 38, 258, 1982.
69. **Owen, D. F. and Weigert, R. G.,** Mutualism between grasses and grazers: an evolutionary hypothesis, *Oikos*, 36, 376, 1981.
70. **Painter, E. L. and Detling, J. K.,** Effects of defoliation on net photosynthesis and regrowth of western wheatgrass, *J. Range Manage.*, 34, 68, 1981.
71. **Parrella, M. P., Jones, V. P., Youngman, R. R., and Lebeck, L. M.,** Effect of leaf mining and leaf stippling of *Liriomyza* spp. on photosynthetic rates of chrysanthemum, *Ann. Entomol. Soc. Am.*, 78, 90, 1985.
72. **Parsons, A. J., Collett, B., and Lewis, J.,** Changes in the structure and physiology of a perennial ryegrass sward when released from a continuous stocking management: implications for the use of exclusion cages in continuously stocked swards, *Grass Forage Sci.*, 39, 1, 1984.
73. **Pedigo, L. P., Hutchins, S. H., and Higley, L. G.,** Economic injury levels in theory and practice, in *Annual Review of Entomology*, Vol. 31, Mittler, T. E., Radovsky, F. J., and Resh, V. H., Eds., Annual Review Inc., Palo Alto, CA, 1986, 341.
74. **Poston, F. L., Pedigo, L. P., Pearce, R. B., and Hammond, R. B.,** Effects of artificial and insect defoliation on soybean net photosynthesis, *J. Econ. Entomol.*, 69, 109, 1976.
75. **Proctor, J. T. A., Bodnar, J. M., Blackburn, W. J., and Watson, R. L.,** Analysis of the effects of the spotted tentiform leaf miner *(Phyllonorycter blancardella)* on the photosynthetic characteristics of apple leaves, *Can. J. Bot.*, 60, 2734, 1982.
76. **Prudhomme, T. I.,** The effect of defoliation history on photosynthetic rates in mountain birch, *Rep. Kevo Subarctic Res. Stn.*, 18, 5, 1982.
77. **Rabbinge, R., Dress, E. M., Van der Graff, M., Verberne, F. C. M., and Elo Wess, A.,** Damage effects of cereal aphids in wheat, *Neth. J. Plant Pathol.*, 87, 217, 1981.
78. **Rhoades, D. F.,** Offensive-defensive interactions between herbivores and plants: their relevance in herbivore population dynamics and ecological theory, *Am. Nat.*, 125, 205, 1985.
79. **Rilling, G. and Steffan, H.,** Experiments on the carbon dioxide fixation and the assimilate import by leaf galls of phylloxera, *Dactylosphaera vitifolii*, on grapevine, *Vitis rupestris* 187G, *Angew. Bot.*, 52, 343, 1978.
80. **Root, R. B.,** Organization of a plant-arthropod association in simple and diverse habitats: the fauna of collards *(Brassica oleracea)*, *Ecol. Monogr.*, 37, 317, 1973.
81. **Ryan, J. D., Johnson, R. C., Eikenbary, R. D., and Dorschner, K. W.,** Drought-greenbug interactions: photosynthesis of greenbug resistant and susceptible wheat, *Crop Sci.*, 27, 283, 1987.
82. **Rule, G. J. A. and Powell, C. E.,** Defoliation and regrowth in the graminaceous plant: the role of current assimilate, *Ann. Bot. (London)*, 39, 297, 1975.
83. **Sances, F. V., Wyman, J. A., and Ting, I. P.,** Morphological responses to spider mite infestation on strawberries, *J. Econ. Entomol.*, 72, 1979.
84. **Sances, F. V., Wyman, J. A., Ting, I. P.,**

Van Steenwyk, R. A., and Oatman, E. R., Spider mite interactions with photosynthesis, transpiration and productivity of strawberry, *Environ. Entomol.,* 10, 442, 1981.

85. **Sances, F. V., Wyman, J. A., and Ting, I. P.,** Physiological responses to spider mite infestation of strawberries, *Environ. Entomol.,* 8, 711, 1979.
86. **Sances, F. V., Toscano, N. C., Hoffmann, M. P., Lapre, L. F., Johnson, M. W., and Bailey, J. B.,** Physiological responses of avocado leaves to avocado brown mite feeding injury, *Environ. Entomol.,* 11, 516, 1982.
87. **Sances, F. V., Toscano, N. C., Oatman, E. R., Lapre, L. F., Johnson, M. W., and Voth, V.,** Reductions in plant processes by *Tetranychus urticae* (Acari Tetranychidae) feeding on strawberry, *Environ. Entomol.,* 11, 733, 1982.
88. **Sanders, T. H., Ashley, D. A., and Brown, R. H.,** Effects of partial defoliation on petiole phloem area, photosynthesis, and ^{14}C translocation in developing soybean leaves, *Crop Sci.,* 17, 548, 1977.
89. **Satoh, M., Kreidemann, P. E., and Loveys, B. R.,** Changes in photosynthetic activity and related processes following decapitation in mulberry trees, *Physiol. Plant.,* 41, 203, 1977.
90. **Schultz, J. C. and Baldwin, I. T.,** Oak leaf quality declines in response to defoliation by gypsy moth larvae, *Science,* 217, 149, 1982.
91. **Sweet, G. B. and Wareing, P. F.,** Role of plant growth in regulating photosynthesis, *Nature (London),* 210, 77, 1966.
92. **Syvertsen, J. P. and McCoy, C. W.,** Leaf feeding injury to citrus by root weevil adults: leaf area, photosynthesis, and water use efficiency, *Fla. Entomol.,* 68, 386, 1985.
93. **Thorne, J. H. and Koller, H. R.,** Influence of assimilate demand on photosynthesis, diffusive resistance, translocation, and carbohydrate levels of soybean leaves, *Plant Physiol.,* 54, 201, 1974.
94. **Tonkyn, D. W. and Whitcomb, R. F.,** Feeding strategies and the guild concept among vascular feeding insects and microorganisms, in *Current Topics in Vector Research,* Vol. 4, Harris, K. F., Ed., Springer-Verlag, New York, 1987, chap. 6.
95. **Trumble, J. T., Ting, I. P., and Bates, L.,** Analysis of physiological growth and yield responses of celery to *Liriomyza trifolii, Entomol. Exp. Appl.,* 38, 15, 1985.
96. **van Roermund, H. J. W., Groot, J. J. R., Rossing, W. A. H., and Rabbinge, R.,** Simulation of aphid damage in winter wheat, *Neth. J. Agric. Sci.,* 34, 488, 1986.
97. **Veen, B. W.,** Photosynthesis and assimilate transport in potato with top-roll disorder caused by the aphid *Macrosiphum euphorbiae, Ann. Appl. Biol.,* 107, 319, 1985.
98. **Walgenbach, J. F. and Wyman, J. A.,** Potato leafhopper (Homoptera Cicadellidae) feeding damage at various potato growth stages, *J. Econ. Entomol.,* 78, 671, 1985.
99. **Walstad, J. D., Nielsen, D. G., and Johnson, N. E.,** Effect of the pine needle scale on photosynthesis of scotch pine, *For. Sci.,* 19, 109, 1973.
100. **Wareing, P. F., Khalifa, M. M., and Treharne, K. J.,** Rate-limiting processes in photosynthesis at saturating light intensities, *Nature (London),* 220, 453, 1968.
101. **Way, M. J. and Cammell, M.,** Aggregation behaviour in relation to food utilization by aphids, in, Animal Populations in Relation to their Food Resources, Proc. 7th British Insect and Fungicide Conf., Watson, A., Ed., 1970, 229.
102. **Welter, S. C., Rosenheim, J., Johnson, M. W., and Mau, R.,** unpublished data, 1987.
103. **Womack, C. L.,** Reduction in photosynthetic and transpiration rates of alfalfa caused by potato leafhopper (Homoptera Cicadellidae) infestations, *J. Econ. Entomol.,* 77, 508, 1984.
104. **Wood, B. W., Tedders, W. L., and Thompson, J. M.,** Feeding influence of 3 pecan aphid species on carbon exchange and phloem integrity of seedling pecan foliage, *J. Am. Soc. Hortic. Sci.,* 110, 393, 1985.
105. **Wood, B. W. and Tedders, W. L.,** Reduced net photosynthesis of leaves from mature pecan trees by three species of pecan aphid, *J. Entomol. Sci.,* 21, 355, 1986.
106. **Woolhouse, H. W.,** The nature of senescence in plants, *Symp. Soc. Exp. Bot.,* 21, 179, 1967.
107. **Wu, A. and Thrower, L. B.,** The physiological association between *Aphis craccivora* Koch and *Vigna sesquipedalis* Fruw, *New Phytol.,* 88, 1981.
108. **Youngman, R. R. and Barnes, M. M.,** Interaction of spider mites (Acari: Tetranychidae) and water stress on gas-exchange rates and water potential of almond leaves, *Environ. Entomol.,* 15, 594, 1986.
109. **Youngman, R. R., Jones, V. P., Welter, S. C., and Barnes, M. M.,** Comparison of feeding damage caused by four tetranychid mite species on gas exchange rates of almond leaves, *Environ. Entomol.,* 15, 190, 1986.

INDEX

A

B

C

H

O

P

Q

R

S

T

U

V

W

X

Y

Z